新时代·新文科×新工科·数字经济高质量人才培养系列（数字产业化）
中国人民大学研究生精品教材建设项目立项资助

Python编程
从数据分析到数据科学
（第2版）

◆ 朝乐门 著

电子工业出版社
Publishing House of Electronics Industry
北京·BEIJING

内 容 简 介

本书是为具有数据思维的数据科学、数据分析和大数据应用人群编写的 Python 学习图书。本书改变了同类图书中普遍存在的"将 Python 当作 C/Java 来教（或学）"的现状，强调了 Python 在数据分析和数据科学中的特殊语法和数据思维；同时，改变了传统图书中"先将知识点、后摆代码"的编写风格，首次将代码放在中心位置，配合最必要的文字介绍，做到主次分明、一目了然，便于学习。

本书主要介绍大数据人才常用的 Python 语言及第三方扩展库的基础知识、思路、方法、经验和技巧，建立了从 Python 到数据分析再到数据科学的通道，形成了 Python 知识、数据分析和数据科学三个知识领域融为一体的知识模式。

本书既可以作为从事数据科学、数据分析和大数据应用人群的入门级系统学习图书，又可以作为相关高校数据科学与大数据技术、大数据应用与管理、信息管理和大数据应用、数据分析、信息分析等专业方向的教材。

未经许可，不得以任何方式复制或抄袭本书之部分或全部内容。
版权所有，侵权必究。

图书在版编目(CIP)数据

Python 编程：从数据分析到数据科学/ 朝乐门著. —2 版. —北京：电子工业出版社，2021.6
ISBN 978-7-121-41200-4

Ⅰ．① P… Ⅱ．① 朝… Ⅲ．① 软件工具—程序设计 Ⅳ．① TP311.561

中国版本图书馆 CIP 数据核字（2021）第 094353 号

责任编辑：章海涛　　文字编辑：刘御廷
印　　刷：北京虎彩文化传播有限公司
装　　订：北京虎彩文化传播有限公司
出版发行：电子工业出版社
　　　　　北京市海淀区万寿路 173 信箱　邮编：100036
开　　本：787×1092　1/16　印张：32.25　字数：820 千字
版　　次：2019 年 1 月第 1 版
　　　　　2021 年 6 月第 2 版
印　　次：2025 年 2 月第 8 次印刷
定　　价：98.00 元

凡所购买电子工业出版社图书有缺损问题，请向购买书店调换。若书店售缺，请与本社发行部联系，联系及邮购电话：（010）88254888，88258888。

质量投诉请发邮件至 zlts@phei.com.cn，盗版侵权举报请发邮件至 dbqq@phei.com.cn。
本书咨询联系方式：192910558（QQ 群）。

第 2 版前言

本书出版后深受广大读者热烈欢迎，成为 Python 与数据科学领域的热门畅销书之一，帮助数万读者顺利掌握从事数据分析和数据科学工作所需的 Python 编程基本功，好评如潮。

为了进一步提升本书质量，我们对本书第 1 版进行了优化，具体说明如下：

（1）**延续**了第 1 版的"简明、实用、高效、优雅"的独特编写思路和良好的阅读与学习体验。

（2）**纠正**了第 1 版中发现的所有拼写错误和表述缺陷。

（3）**采纳**了全国读者对第 1 版的修改建议和问题反馈。

（4）**新增**了两章内容，即"45 人脸识别与图像分析"和"48 基于 Spark 和 MongoDB 的大数据分析"。

（5）**删除**了两章内容，即第 1 版的"46 Spark Python 开发环境的搭建"和"47 NoSQL 数据库"。

（6）**调整**了两章内容的顺序，即第 1 版中的"20 迭器与生成器"和"41 自然语言处理"，分别调整为第 2 版的"25 迭代器与生成器"和"44 自然语言处理"。

（7）**补充**了部分新知识点，如模型调参、基于 Yellowbrick 的机器学习可视化、基于 OpenCV 的图像处理、基于 Spark+MongoDB+MLib 的大数据分析等。

（8）**更新**了配套 PPT、习题库、教学大纲等教学资源。

（9）**优化**了阅读体验，如"参见本书……"语句中注明目标内容所在的"页码+行号"以及页眉中增加章节名称。

（10）**提供**了在课堂教学中学生经常提出的疑难问题及其解答。

本书凝聚了作者及其团队、来自全国各地高校相关课程的广大师生以及 Python 与数据科学爱好者的共同努力。在此特别感谢：一直以来关心与支持本书的每一位读者；我的博士、硕士研究生肖纪文、孙智中和张晨，他（她）们参与并完成第 2 版校对工作；华北电力大学控制与计算机工程学院副院长师瑞峰教授及其学生们，他们对本书第 1 版给出了详尽的教材使用评价与反馈报告；电子工业出版社的所有编辑及各位领导，他们为本书的出版做了大量细致工作。本书第 2 版受 2021 年度中国人民大学研究生精品教材建设项目立项资助，第 1 版受教育部-IBM 产学研合作协同育人项目支持。

Good luck with Python!

朝乐门
于中国人民大学
2021 年 6 月

第 1 版前言

"写好一本书"是一件功德无量的事情。因为,写书可以用自己的时间来节约他人的时间。在这浮躁而现实的时代,不去写论文或争课题,而是低头写教材,明知不会计入工作量或业绩,但我还是乐此不疲。按自己的常规速度计算,写一本教材大概也就 3 个月的时间,但这次实际投入了整整 18 个月的精力,目的只有一个——用自己的 18 个月,为他人节约 15 个月的时间。

"写好一本书"需要有好的顶层设计。我在近几年的教学一线中意识到,国内亟需一本面向数据科学和大数据技术专业人才培养的 Python 好教材。目前,相关图书中存在的问题有:第一,"将 Python 当作 Java/C 来教(或学)",换一个"新语言"来讨论"老问题",根本品不出 Python 特有的味道;第二,"先讲知识点,后摆代码"式教材编写风格,以"文字"为主,"代码"为辅,导致主次颠倒;第三,"数据科学类专业与计算机科学类专业中选用的 Python 教材没什么区别",不知道谁是谁;第四,"把读者(或自己)当作编程小白来写(或读)"。目前,多数读者已有 C/Java 等语言的基础,Python 属于"第二外语",不需要低级重复,更不应该用不同语言反复学习同一个知识点,始终徘徊在知识殿堂的门口。突破上述局限并且探索出新的教学模式和教材编写方式是我编写这本书的初心。是否达到了预期目标,需要各位仔细阅读全书之后明鉴。

"写好一本书"需要有 10 本书的真材实料。在本书撰写过程中,我参阅了大量国内外专著、教材、论文、开源项目和原始数据,虽然书中对参考文献多有标注,但难免挂一漏万,希望相关作者见谅。书中还借鉴了自 2015 年以来我和我的团队撰写或翻译的部分短文,同时借鉴了我的课堂上学生曾提问或关注的问题。

"写好一本书"离不开他人的鼎力相助。电子工业出版社领导及编辑,尤其是章海涛编辑为本书的出版做了大量工作;感谢教育部—IBM 产学合作协同育人项目的资助和支持;感谢中国人民大学刘岩、杨灿军、李昊璟、王雨晴、曲涵晴、赵群、李雪明、张晨、冀佳钰等学生参与了本书的校对工作;感谢家人给予了长期的理解与支持,本人从事基础研究,淡泊名利,她们却从不抱怨。

"写好一本书"是一个长期反复打磨的过程。本书必有不足之处,望各位读者不吝赐教。我们将通过华信教育资源网(http://www.hxedu.com.cn)和微信公众号"数据科学 DataScience"实时发布勘误或更新内容。这是继《数据科学》《数据科学理论与实践》之后我的第三本书。曾有人跟我说,"朝老师,您硕果累累,还那么拼命,未来一定是数据科学领域最大的牛。"我回答,"No,那不是我的目的。我的奋斗目的只有一个:争取做到数据科学领域最努力的人,也就是最舍得投入时间和精力的人,其他的都无所谓。"希望"我的努力"成为"您的努力"!

朝乐门
于中国人民大学
2018 年 12 月

目　录

第一篇　准备工作

1 为什么要学习 Python，学习 Python 的什么 ... 3
2 学习 Python 之前需要准备的工作 ... 6
3 如何看懂和运行本书代码 ... 8
   3.1 输入部分 ... 8
   3.2 输出部分 ... 10
   3.3 错误与异常信息 ... 11
   3.4 外部数据文件 ... 12
   3.5 注意事项 ... 14

第二篇　Python 基础

4 数据类型 ... 19
   4.1 查看数据类型的方法 ... 20
   4.2 判断数据类型的方法 ... 21
   4.3 转换数据类型的方法 ... 22
   4.4 特殊数据类型 ... 23
   4.5 序列类型 ... 26
5 变量 ... 28
   5.1 变量的定义方法 ... 29
   5.2 Python 是动态类型语言 ... 29
   5.3 Python 是强类型语言 ... 30
   5.4 Python 中的变量名是引用 ... 31
   5.5 Python 中区分大小写 ... 32
   5.6 变量命名规范 ... 32
   5.7 iPython 的特殊变量 ... 33
   5.8 查看 Python 关键字的方法 ... 34
   5.9 查看已定义的所有变量 ... 35
   5.10 删除变量 ... 37
6 语句书写规范 ... 39
   6.1 一行一句 ... 40
   6.2 一行多句 ... 40

目　录

6.3　一句多行 ·· 41

6.4　复合语句 ·· 42

6.5　空语句 ·· 43

7　赋值语句 ··· 44

7.1　赋值语句在 Python 中的重要地位 ··· 45

7.2　链式赋值语句 ·· 45

7.3　复合赋值语句 ·· 46

7.4　序列的拆包式赋值 ·· 46

7.5　两个变量值的调换 ·· 47

8　注释语句 ··· 48

8.1　注释方法 ·· 48

8.2　注意事项 ·· 49

9　运算符 ··· 50

9.1　特殊运算符 ·· 53

9.2　内置函数 ·· 57

9.3　math 模块 ··· 58

9.4　优先级与结合方向 ·· 59

10　if 语句 ··· 61

10.1　基本语法 ·· 61

10.2　elif 语句 ··· 62

10.3　if 与三元运算 ·· 63

10.4　注意事项 ·· 64

11　for 语句 ·· 66

11.1　基本语法 ·· 67

11.2　range()函数 ·· 67

11.3　注意事项 ·· 68

12　while 语句 ··· 71

12.1　基本语法 ·· 71

12.2　注意事项 ·· 72

13　pass 语句 ··· 74

13.1　含义 ·· 74

13.2　作用 ·· 75

14　列表 ··· 76

14.1　定义方法 ·· 78

14.2 切片操作	79
14.3 反向遍历	81
14.4 类型转换	83
14.5 extend 与 append 的区别	83
14.6 列表推导式	84
14.7 插入与删除	87
14.8 常用操作函数	89
15 元组	**94**
15.1 定义方法	95
15.2 主要特征	97
15.3 基本用法	99
15.4 应用场景	100
16 字符串	**103**
16.1 定义方法	104
16.2 主要特征	105
16.3 字符串的操作	106
17 序列	**111**
17.1 支持索引	112
17.2 支持切片	113
17.3 支持迭代	114
17.4 支持拆包	114
17.5 支持*运算	115
17.6 通用函数	117
18 集合	**120**
18.1 定义方法	121
18.2 主要特征	122
18.3 基本运算	123
18.4 应用场景	125
19 字典	**126**
19.1 定义方法	127
19.2 字典的主要特征	128
19.3 字典的应用场景	129
20 函数	**130**
20.1 内置函数	131

目　录

　　20.2　模块函数 131
　　20.3　用户自定义函数 132
21　内置函数 133
　　21.1　内置函数的主要特点 134
　　21.2　数学函数 134
　　21.3　类型函数 135
　　21.4　其他功能函数 136
22　模块函数 141
　　22.1　import 模块名 142
　　22.2　import 模块名 as 别名 143
　　22.3　from 模块名 import 函数名 143
23　自定义函数 145
　　23.1　定义方法 147
　　23.2　函数中的 docString 148
　　23.3　调用方法 148
　　23.4　返回值 149
　　23.5　形参与实参 150
　　23.6　变量的可见性 152
　　23.7　值传递与地址传递 154
　　23.8　其他注意事项 156
24　lambda 函数 158
　　24.1　lambda 函数的定义方法 159
　　24.2　lambda 函数的调用方法 160

第三篇　Python 进阶

25　迭代器与生成器 165
　　25.1　可迭代对象与迭代器 166
　　25.2　生成器与迭代器 167
26　模块 169
　　26.1　导入与调用用法 170
　　26.2　查看内置模块清单的方法 171
27　包 174
　　27.1　相关术语 175
　　27.2　安装包 175

27.3 查看已安装包 ... 176
27.4 更新（或删除）已安装包 ... 176
27.5 导入包 ... 177
27.6 查看包的帮助 ... 178
27.7 常用包 ... 179

28 帮助文档 ... 180
28.1 help 函数 ... 181
28.2 docString ... 181
28.3 查看源代码 ... 182
28.4 doc 属性 ... 183
28.5 dir()函数 ... 184
28.6 其他方法 ... 186

29 异常与错误 ... 187
29.1 try/except/finally ... 188
29.2 异常信息的显示模式 ... 189
29.3 断言 ... 190

30 程序调试方法 ... 192
30.1 调试程序的基本方法 ... 193
30.2 设置错误信息的显示方式 ... 194
30.3 设置断言的方法 ... 195

31 面向对象编程 ... 197
31.1 类的定义方法 ... 198
31.2 类中的特殊方法 ... 199
31.3 类之间的继承关系 ... 201
31.4 私有属性及@property 装饰器 ... 203
31.5 self 和 cls ... 204
31.6 new 与 init 的区别和联系 ... 205

32 魔术命令 ... 208
32.1 运行.py 文件：%run ... 209
32.2 统计运行时间：%timeit 与%%timeit ... 210
32.3 查看历史 In 和 Out 变量：%history ... 211
32.4 更改异常信息的显示模式：%xmode ... 212
32.5 调试程序：%debug ... 214
32.6 程序运行的逐行统计：%prun 与%lprun ... 215

目　录

32.7 内存使用情况的统计：%memit ··· 216
33 搜索路径 ··· 218
33.1 变量搜索路径 ··· 219
33.2 模块搜索路径 ··· 221
34 当前工作目录 ··· 224
34.1 显示当前工作目录的方法 ··· 225
34.2 更改当前工作目录的方法 ··· 225
34.3 读、写当前工作目录的方法 ··· 226

第四篇　数据加工

35 随机数 ··· 229
35.1 一次生成一个数 ··· 230
35.2 一次生成一个随机数组 ··· 231
36 数组 ··· 234
36.1 创建方法 ··· 238
36.2 主要特征 ··· 241
36.3 切片/读取 ··· 243
36.4 浅拷贝和深拷贝 ··· 249
36.5 形状和重构 ··· 250
36.6 属性计算 ··· 254
36.7 ndarray 的计算 ··· 256
36.8 ndarray 的元素类型 ··· 258
36.9 插入与删除 ··· 259
36.10 缺失值处理 ··· 260
36.11 ndarray 的广播规则 ··· 261
36.12 ndarray 的排序 ··· 262
37 Series ··· 265
37.1 Series 的主要特点 ··· 266
37.2 Series 的定义方法 ··· 266
37.3 Series 的操作方法 ··· 269
38 DataFrame ··· 274
38.1 DataFrame 的创建方法 ··· 277
38.2 查看行或列 ··· 278
38.3 引用行或列 ··· 279

- 38.4 index 操作 ... 282
- 38.5 删除或过滤行/列 ... 284
- 38.6 算术运算 ... 289
- 38.7 大小比较运算 ... 295
- 38.8 统计信息 ... 296
- 38.9 排序 ... 298
- 38.10 导入/导出 ... 300
- 38.11 缺失数据处理 ... 301
- 38.12 分组统计 ... 307

39 日期与时间 ... 310
- 39.1 常用包与模块 ... 311
- 39.2 时间和日期类型的定义 ... 311
- 39.3 转换方法 ... 313
- 39.4 显示系统当前时间 ... 315
- 39.5 计算时差 ... 316
- 39.6 时间索引 ... 316
- 39.7 period_range()函数 ... 319

40 可视化 ... 320
- 40.1 Matplotlib 可视化 ... 322
- 40.2 改变图的属性 ... 325
- 40.3 改变图的类型 ... 328
- 40.4 改变图的坐标轴的取值范围 ... 329
- 40.5 去掉边界的空白 ... 332
- 40.6 在同一个坐标上画两个图 ... 332
- 40.7 多图显示 ... 333
- 40.8 图的保存 ... 334
- 40.9 散点图的画法 ... 335
- 40.10 Pandas 可视化 ... 336
- 40.11 Seaborn 可视化 ... 339
- 40.12 数据可视化实战 ... 342

41 Web 爬取 ... 345
- 41.1 Scrapy 的下载与安装 ... 348
- 41.2 Scrapy Shell 的基本原理 ... 349
- 41.3 Scrapy Shell 的应用 ... 350

目　录

　　41.4　自定义 Spider 类 ··· 352
　　41.5　综合运用 ·· 359

第五篇　数据分析

42　统计分析 ·· 367
　　42.1　业务理解 ·· 369
　　42.2　数据读入 ·· 369
　　42.3　数据理解 ·· 370
　　42.4　数据准备 ·· 371
　　42.5　模型类型的选择与超级参数的设置 ·· 373
　　42.6　训练具体模型及查看其统计量 ·· 374
　　42.7　拟合优度评价 ·· 376
　　42.8　建模前提假定的讨论 ·· 376
　　42.9　模型的优化与重新选择 ·· 378
　　42.10　模型的应用 ·· 382

43　机器学习 ·· 383
　　43.1　机器学习的业务理解 ·· 384
　　43.2　数据读入 ·· 385
　　43.3　数据理解 ·· 386
　　43.4　数据准备 ·· 389
　　43.5　算法选择及其超级参数的设置 ·· 392
　　43.6　具体模型的训练 ·· 393
　　43.7　用模型进行预测 ·· 393
　　43.8　模型评价 ·· 394
　　43.9　模型的应用与优化 ··· 395

44　自然语言处理 ·· 399
　　44.1　常用包 ·· 400
　　44.2　自然语言处理包的导入及设置 ·· 400
　　44.3　数据读入 ·· 401
　　44.4　分词处理 ·· 403
　　44.5　自定义词汇 ·· 404
　　44.6　停用词处理 ·· 409
　　44.7　词性分布分析 ·· 410
　　44.8　高频词分析 ·· 412

44.9 词频统计 ··· 414
44.10 关键词分析 ·· 415
44.11 生成词云 ·· 416

45 人脸识别与图像分析 419
45.1 安装并导入 opencv-python 包 ··· 420
45.2 读取图像文件 ·· 420
45.3 将 RGB 图像转换为灰度图 ·· 420
45.4 人脸识别与绘制长方形 ·· 421
45.5 图像显示 ··· 422
45.6 图像保存 ··· 423

第六篇 大数据处理

46 Spark 编程 425
46.1 导入 pyspark 包 ·· 427
46.2 SparkSession 及其创建 ··· 427
46.3 Spark 数据抽象类型 ·· 430
46.4 Spark DataFrame 操作 ·· 433
46.5 SQL 编程 ··· 437
46.6 DataFrame 的可视化 ·· 440
46.7 Spark 机器学习 ··· 442
    46.7.1 创建 Spark Session ·· 443
    46.7.2 读入数据 ·· 443
    46.7.3 数据理解 ·· 444
    46.7.4 数据准备 ·· 444
    46.7.5 模型训练 ·· 446
    46.7.6 模型评价 ·· 446
    46.7.7 预测 ··· 447

47 基于 Spark 和 MongoDB 的大数据分析 449
47.1 数据准备 ··· 450
47.2 数据读入 ··· 452
47.3 数据理解 ··· 453
47.4 数据准备 ··· 455
47.5 模型训练 ··· 458

目　录

47.6　模型评价 ··· 459
47.7　模型应用 ··· 461

第七篇　继续学习

48　Python 初学者常见错误及纠正方法 ··· 465
48.1　NameError: name 'xxxx' is not defined ······································· 465
48.2　IndentationError: unexpected indent ··· 466
48.3　SyntaxError: invalid character in identifier ······························· 467
48.4　TypeError: 'XXXX' object does not support item assignment ··············· 468
48.5　TypeError: unsupported operand type(s) for XXXX ·························· 468
48.6　IndexError: list index out of range ·· 469
48.7　TypeError: type() takes XXXX arguments ····································· 470
48.8　SyntaxError: unexpected EOF while parsing ·································· 470
48.9　ModuleNotFoundError: No module named XXXX ································· 471
48.10　TypeError: 'list' object is not callable ·································· 472
48.11　SyntaxError: invalid syntax ·· 473
48.12　AttributeError: XXXX object has no attribute XXXX ························ 474
48.13　TypeError: XXXX object is not an iterator ································· 475
48.14　FileNotFoundError: File XXXX does not exist ······························ 476
48.15　IndexError: too many indices for array ···································· 478
48.16　TypeError: Required argument XXXX not found ······························ 479
48.17　TypeError: an XXXX is required (got type YYYY) ··························· 480
48.18　ValueError: Wrong number of items passed XXXX, placement
implies YYYY ··· 481

49　Python 数据分析和数据科学面试题 ··· 483
50　继续学习本书内容的推荐资源 ··· 494
50.1　重要网站 ··· 494
50.2　重要图书 ··· 494
50.3　常用模块与工具包 ··· 495
50.4　常用统计模型 ·· 495
50.5　核心机器学习算法 ··· 496
50.6　继续学习数据科学的建议路线图 ··· 497

参考文献 ··· 501

第一篇
准备工作

为什么要学习 Python
学习 Python 的什么
学习 Python 之前需要准备的工作
如何看懂和运行本书代码

1 为什么要学习 Python，学习 Python 的什么

 常见疑问及解答。

Life is short, you need Python.（人生苦短，你需要 Python。见图1-1）

首先，Python有哪些优点？

- Python的设计哲学：优雅、明确、简练
- Python的设计目的：符合数据分析与数据科学项目需要
- Python已成为数据分析和数据科学领域最受欢迎的语言

但是，我已经学过C、Java、C++等语言，为什么还要需要学习Python？

- 有三个层面的原因，见图1-2

那么，我应该学习Python的哪些知识？

- 通常说"Python is everything"，意思就是"可以用Python来做 Everything"，如桌面应用、Web程序、移动App等

因此

- 我们只需要学习"面向数据分析与数据科学的Python"，而不是"Python的全部知识"，见图1-3～图1-5

1 为什么要学习 Python，学习 Python 的什么

图 1-1 Python 之父 Guido van Rossum 与 Python 官网

第一层原因

最直接的原因，可以说是"表层原因"。Java、C等语言是为软件开发而设计的，不适合做数据科学任务

比如，数据集的读写和排序是数据科学中经常处理的工作，如果用Java写的话，需要多层for语句的代码，很麻烦。但是，在Python或R语言中，这些问题变得很简单——支持向量化计算，可以直接读写数据集（不需要for语句）；Python和R语言采用泛型函数式编程，可以直接调用函数 sorted/sort()来实现数据集的排序工作（不需要你自己写排序算法和代码）

因此，如果坚持用Java、C等语言完成数据科学任务，精力将消耗在流程控制、数据结构的定义和算法设计上，而不是在集中在处理数据问题上

第二层原因

我们可以通过 Python/R语言调用面向数据科学任务的专业级服务——包/模块

目前，PyPI（Python Package Index）上可提供的Python模块或包已经超过13940个。CRAN上可用的R包至少有10381个。也就是说，Python/R语言本身并不是奥妙所在，而是Python/R语言背后的包的功能非常强大。例如，用Java、C等语言实现数据可视化非常复杂，而且不美观，而用Python的Matplotlib或R的ggplot2即可轻松实现

因此，我们用 Python/R语言并不是因为它们本身比Java、C等语言更强大，而是因为Python/R语言可以调用众多专门用于数据分析与数据科学任务的包

第三层原因

其实，第二层原因也不是根本原因，根本原因是Python/R语言的背后，尤其是主流Python/R包的开发者都是统计学、机器学等数据科学领域的大牛

Python包Pandas的开发者Wes McKinney是Python领域的最活跃的科学家之一，著有数据科学领域的重要图书《Python for Data Analysis: Data Wrangling with Pandas, NumPy, and IPython》等。R包ggplot2的开发者Hadley Wickham是R领域最活跃的数据科学家之一，著有数据科学领域的重要图书《R for Data Science: Visualize, Model, Transform, Tidy, and Import Data》

因此，如果用 Python/R语言，你就找到了组织，找到了同类——与世界顶级的数据科学家们站在一起，利用他们的思想指导自己，利用他们的智慧解决自己的数据科学问题，这才是根本原因

图 1-2 学习 Python 的三层原因

图 1-3　Wes McKinney 及其著名图书《Python for Data Analysis》

图 1-4　Hadley Wickham 及其著名图书《R for Data Science》

图 1-5　Sandy Ryza 及其著名图书《Advanced Analytics with Spark》

提示

iPython：Python 的一种交互式（interactive）的 Shell（编程/计算环境），是 Jupyter 的内核。因此，数据科学和数据分析项目中广泛采用的 Jupyter Notebook（原名 Ipython Notebook）具备 iPython 的诸多特征和功能，详见 https://ipython.org。

2 学习 Python 之前需要准备的工作

 常见疑问及解答。

建议安装 Anaconda，因为编写 Python 代码时会遇到如下问题。

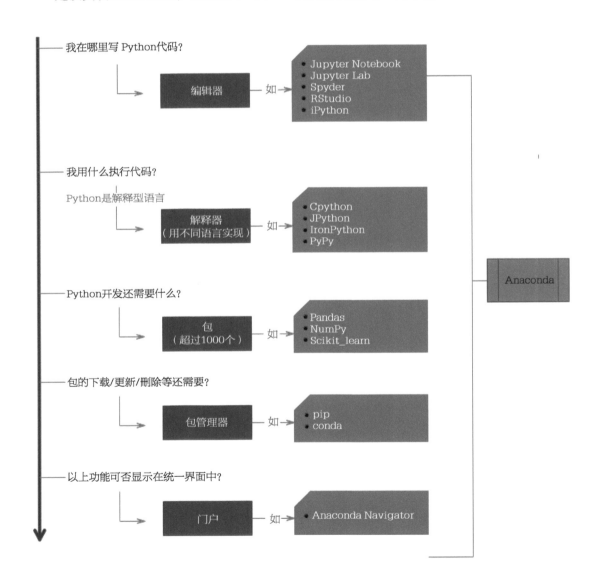

提示 如果已安装了 Anaconda，就不需要逐个下载或安装 Python 编程所需的编辑器、解释器、包及包管理器，也不需要手动配置它们的关联参数。

注意 本书采用的是数据科学领域最常用的编辑器 Jupyter Notebook/Lab 和内核（解释器）Python 3。

提示 Anaconda 的下载和安装方法参见 Anaconda 官网。

Anaconda 官网及简介如图 2-1 所示。安装 Anaconda 后的 Windows "开始"菜单项及其用法如图 2-2 所示。

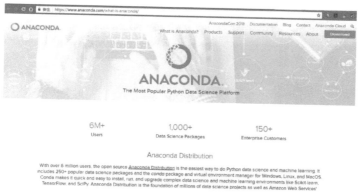

图 2-1　Anaconda 官网及简介

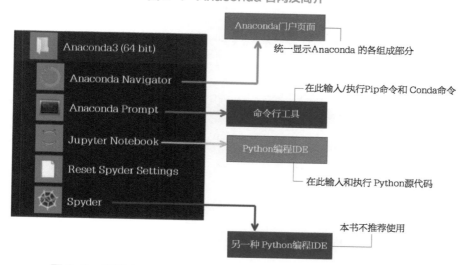

图 2-2　安装 Anaconda 后的 Windows "开始"菜单项及其用法

2 学习 Python 之前需要准备的工作

Jupyter Notebook/Lab 的编辑器用法如图 2-3 所示。

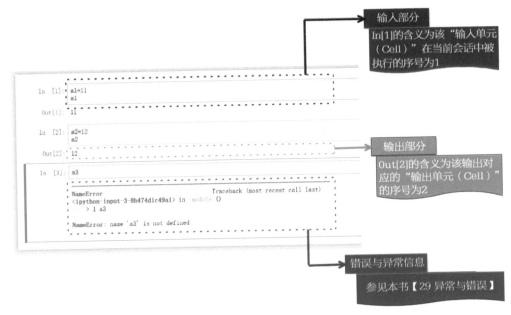

图 2-3 Jupyter Notebook/Lab 的编辑器用法

3 如何看懂和运行本书代码

3.1 输入部分

In [1]:

提示 读者只需要输入在 Jupyter Notebook 的输入变量 "In[]:" 右侧的 "单元（Cell）" 中的内容。

```
x1 = 11
x1
```

注意 Python 中需要注意区分大小写以及注意代码行的对齐方式（代码缩进），详见本书【6.4 复合语句】。

技巧 在 Jupyter Notebook 中运行一个 "单元（Cell）" 的默认快捷键为 Ctrl+Enter，更多快捷键见 Jupyter Notebook 的菜单栏中的 Help/ Keyboard Shortcuts 菜单项。

In [2]:

```
#定义变量 x2
x2 = 12
x2
```

提示 在 "输入单元（Cell）" 中，凡是以 "#" 开始的行均为 Python 注释语句，读者可以不用输入，详见本书【8 注释语句】。

In [3]:

```
x3 = 13
x3
```

提示 In[]中的编号为该 "输入单元（Cell）" 在 "Jupyter Notebook Kernel 中的当前会话（Session）" 中被执行的顺序号。

3 如何看懂和运行本书代码

> **注意** 同一个"单元（Cell）"被多次执行时，其"In[]变量"的编号会随之发生变化，详见本书【3.5 注意事项】。
>
> **技巧** 强制停止和重新启动"当前会话（Session）"的方法分别为菜单栏中的 Kernel/Interrupt 和 Kernel/Restart。

In [4]:
```
x4 = 14
```

In [5]:
```
x4 = x4+1
x4
```

> **提示** "x4 = x4+1"是一种自我赋值语句，每执行一次该"单元（Cell）"，x4 的值都会增 1。
>
> **思路** 在数据分析和数据科学项目中，跟踪和查看"变量的当前值"尤为重要。

In [6]:
```
x5 = 5
x5
```

> **思路** 在 Jupyter Notebook 中，Python 代码的执行是以"单元（Cell）"为单位进行的，执行顺序与 C/Java 语言不一样，并非按预定顺序（如从上到下）逐条执行。
>
> **提示** Python 中没有 main() 函数，"单元（Cell）"的执行顺序由用户指定，与其位置无关。

Out [6]:   5

3.2 输出部分

In [7]:

 提示 输出部分显示在 Jupyter Notebook 的输出变量 "Out[]:" 右侧的 "单元（Cell）" 中。

y1 = 21
y1

 注意 Jupyter Notebook 中的输出以 "单元（Cell）" 为单位，输出结果紧跟在对应的 "输入单元（Cell）" 下方。

Out [7]:   21

In [8]:

y2 = 22
y2

 提示 Out[]中的编号为该输出结果所对应的 In[]编号。

Out [8]:   22

In [9]:

y3 = 23
print(y3)

 注意 在 Python 中，用 print()函数输出时不会放入 Out 变量，即输出结果中没有 Out[]编号。

Out [9]:   23

3 如何看懂和运行本书代码

> **思考** In[8]和In[9]中的y2和print(y3)均可以输出结果，二者有区别吗？

有。主要区别有3个：① 前者并不是Python的语法，而是Jupyter Notebook为了方便我们编程而提供的功能。在Python中，标准输出仍需用print()函数；② 前者是Jupyter Notebook的语法，会将输出结果放入Jupyter Notebook的Out队列变量，且有Out[]编号；后者不会被放入Jupyter Notebook的Out队列变量，且无Out[]编号；③ 前者是Jupyter Notebook进行"优化"后的显示结果，在输出效果上，往往与print()函数有所差别。

3.3 错误与异常信息

In [10]:
```
z1 = 31
z
```

> **注意** 报错信息为"z为未定义对象"，原因是定义时使用的变量名并非为"z"，而是"z1"。

> **提示** Python解释器的报错/异常信息，参见本书【48 Python初学者常见错误及纠正方法（P467/In[1]）】和【29 异常与错误（P189/In[3]）】

Out [10]: ------------------------------------
```
NameError                                 Traceback (most recent call last)
<ipython-input-10-8d66e1a13261> in <module>()
      1 z1=31
----> 2 z

NameError: name 'z' is not defined
```

3.4 外部数据文件

In [11]: 提示 | 对外部数据文件（如 Excel、CSV、JSON 等）进行编程之前，需要事先将其放在"当前工作目录"下。

 技巧 | 显示"当前工作目录"的方法如下。

```
import os
print(os.getcwd())
```

 提示 | 修改当前工作目录及更多内容参见本书【34 当前工作目录】（P225/In[3]）。

Out [11]: C:\Users\soloman\clm

In [12]:
```
from pandas import read_csv
data = read_csv('bc_data.csv')
data.head(2)
```

 注意 | 需要将数据文件（bc_data.csv）事先放在"当前工作目录"（C:\Users\soloman\clm）下，否则报 FileNotFoundError 错误。

 提示 | 该"单元（Cell）"中的代码含义为，用 Pandas 的 read 编码读取外部文件"bc_data.csv"。参见本书【38 DataFrame（P301/In[74]）】。

Out [12]:

	id	diagnosis	radius_mean	texture_mean	perimeter_mean	area_mean	smoothne
0	842302	M	17.99	10.38	122.8	1001.0	0.11840
1	842517	M	20.57	17.77	132.9	1326.0	0.08474

2 rows × 32 columns

3 如何看懂和运行本书代码

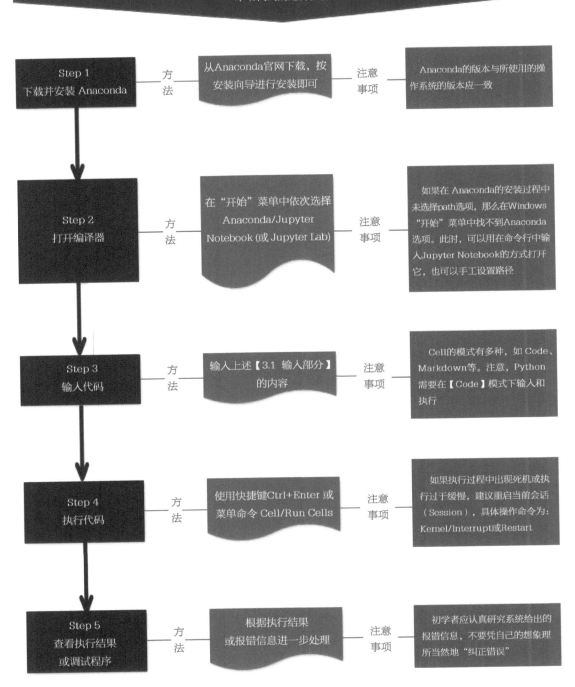

3.5 注意事项

Python 编程中的七个注意事项如图 3-1 所示。

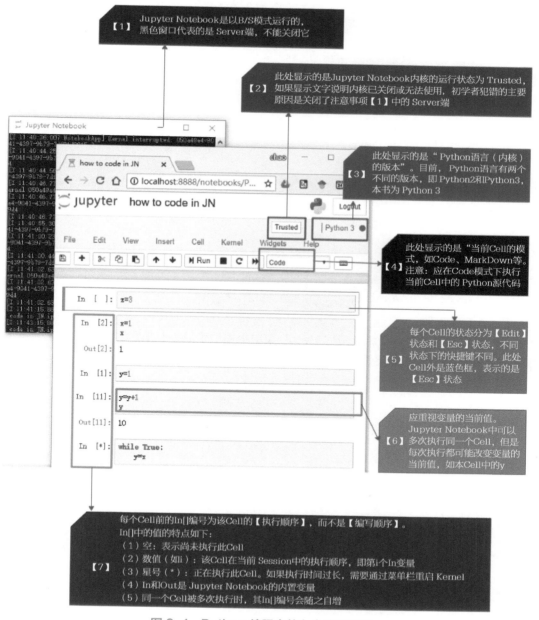

图 3-1　Python 编程中的七个注意事项

3 如何看懂和运行本书代码

Jupyter Notebook 中 Cell 的 Esc 状态和 Edit 状态如图 3-2 所示。

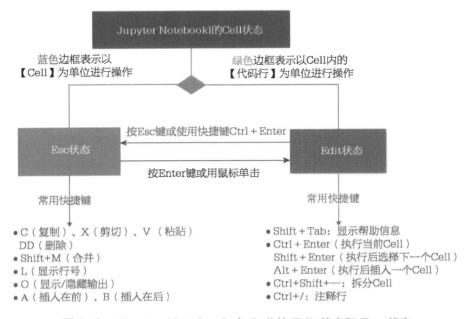

图 3-2 Jupyter Notebook 中 Cell 的 Edit 状态和 Esc 状态

 充分利用 Jupyter Notebook 的 "Help" 菜单。

在 Jupyter Notebook/Lab 的 "Help" 菜单中提供了很多非常好的学习和参考资料，建议初学者充分利用这些重要资源，如图 3-3 所示。

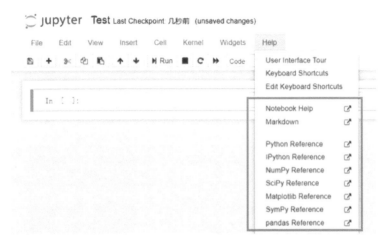

图 3-3 "Help" 菜单

第二篇

Python 基础

- 数据类型
- 变量
- 语句书写规范
- 赋值语句
- 注释语句
- 运算符
- if 语句
- for 语句
- while 语句
- pass 语句
- 列表
- 元组
- 字符串
- 序列
- 集合
- 字典
- 函数
- 内置函数
- 模块函数
- 自定义函数
- lambda 函数

4 数据类型

常见疑问及解答。

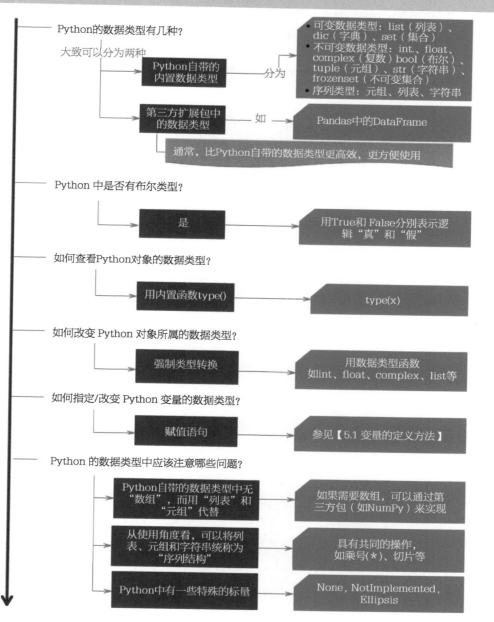

4 数据类型

4.1 查看数据类型的方法

In [1]:
```
#整型(int)
type(1)
```

 可以用 Python 内置函数 type() 查看数据类型。

Out [1]: int

In [2]:
```
#浮点型(float)
type(1.2)
```

Out [2]: float

In [3]:
```
#布尔类型(bool)
type(True)
```

 在 Python 中，用 True 和 False 分别表示逻辑"真"和"假"。

Out [3]: bool

In [4]:
```
#字符串(str)
type("DataScience")
```

 在 Python 中，双引号或单引号是字符串存在的标志，参见本书【16 字符串（P104/In[1]）】。

Out [4]: str

In [5]:
```
#列表(list)
type([1,2,3,4,5,6,7,8,9])
```

 在 Python 中，列表用中括号"[ ]"表示，参见本书【14 列表（P78/In[1]）】。

与 C 和 Java 语言不同的是，Python 语言内置数据类型中没有提供数组类型，而是用"列表"和"元组"代替类似其他语言中的"数组"。

Out[5]:　　list

In [6]:
```
#元组(tuple)
type((1,2,3,4,5,6,7,8,9))
```

 在 Python 中，用圆括号或逗号表示元组，参见本书【15 元组（P95/In[1]）】。元组的英文名称为 tuple。

 需要记住数据类型对应的 Python 关键字。

Out [6]:　　tuple

In [7]:
```
#集合(set)
type({1,2,3,4,5,6,7,8,9})
```

 在 Python 中，用花括号表示集合或字典，参见本书【18 集合（P121/In[1]）】。

Out [7]:　　set

In [8]:
```
#字典(dict)
type({"a":0,"b":1,"c":2})
```

 字典与集合的区别：字典是带 Key 的集合，参见本书【19 字典（P127/In[1]）】。

Out [8]:　　dict

4.2 判断数据类型的方法

In [9]:
```
#判断x是否为整型(int)
x = 10
```

4 数据类型

isinstance(x,int)

思路 | 用内置函数 isinstance()判断变量是否为整型。

Out [9]: True

In [10]:
```
#判断 y 是否为整型(int)
y=10.0
isinstance(y,int)
```

提示 | isinstance()函数的第一个参数 y 为变量名，第二个参数 int 为数据类型的名称。

Out [10]: False

In [11]:
```
#特殊情况
isinstance(True,int)
```

注意 | 输出结果为 True，原因：在 Python 中，bool 类型为 int 类型的子类。

Out [11]: True

4.3 转换数据类型的方法

In [12]:

提示 | 具体方法：进行强制类型转换，如将浮点型转换为整型。

```
int(1.6)
```

Out [12]: 1

In [13]:
```
#将整型转换为浮点型
float(1)
```

>
> **思路** 通常，强制类型转换函数名与目标数据类型的名称一致。例如，需要转换成 float 类型，强制类型转换函数名为 float()。

Out [13]:   1.0

In [14]:
```
#将整型转换为布尔（bool）类型
bool(0)
```

Out [14]:   False

In [15]:
```
#将列表转换为元组（tuple）
tuple([1,2,1,1,3])
```

Out [15]:   (1, 2, 1, 1, 3)

In [16]:
```
#将元组转换为列表（list）
list((1,2,3,4))
```

>
> **提示** 列表（list）和元组（tuple）的区别在于，前者为"可变对象"，后者为"不可变对象"，参见【14 列表（P76）】和【15 元组（P94）】。

Out [16]:   [1, 2, 3, 4]

4.4 特殊数据类型

In [17]:
>
> **提示** 与 C、Java 语言不同的是，Python 语言中有特殊的数据类型。
>
> ```
> #None
> x = None
> print(x)
> ```
>
> **注意** None 的输出必须用 print()函数，否则什么也看不见。

Out [17]:   None

4 数据类型

In [18]:
```
#不支持此类对象
NotImplemented
```

与 R 语言不同的是，Python 语言中支持"标量"的概念，如 Ellipsis、NotImplemented 等。

R 语言中没有"标量"的概念，默认数据类型为"向量"。

Out [18]: NotImplemented

In [19]:
```
#省略号
Ellipsis
```

Out [19]: Ellipsis

In [20]:
```
#complex 类型（复数类型）
x = 2+3j
print('x = ', x)
```

Python 支持复数类型。

Out [20]: x = (2+3j)

In [21]:
```
y=complex(3,4)
```

3+4j 等价于 complex(3,4)。

```
print('y = ', y)
```

Out [21]: y = (3+4j)

In [22]:
```
print('x+y = ', x+y)
```

关于 print()函数的参数及用法，可以通过"print?"或"?print"查看其说明文档。

Out [22]: x+y = (5+7j)

In [23]:
```
#bool 函数与布尔值
bool(1)
```

Out [23]: True

In [24]:
```
bool(0)
```

Out [24]: False

In [25]:
```
bool("abc")
```

Out [25]: True

In [26]:
```
#如何表达"进制"？以二进制为例：
int('100', base = 2)
```

 注意 第一个参数必须用引号括起来。注意：如果没有双引号或单引号，Python 解释器将按变量名来处理，而 Python 变量名需要先定义、后使用。

Out [26]: 4

In [27]:
```
#十进制
int('100', base = 10)
```

 提示 意为十进制（base = 10）数 100。

Out [27]: 100

In [28]:
```
#科学记数法
9.8e2
```

 注意 此处，e 代表的是科学记数法中的 10，而不是自然常数 e = 2.71828。

Out [28]: 980.0

4.5 序列类型

In [29]:
```
mySeq1 = "Data Science"
mySeq2 = [1,2,3,4,5]
mySeq3 = (11,12,13,14,15)
```

 在Python中，"序列"（Sequence）并不是特指一个独立数据类型，而是泛指一种有序的容器，即容器中的元素有先后顺序，即有"下标"的概念。

 Python中常见的序列类型有字符串、列表和元组。但是，集合（set）类型不属于序列，因为集合中的元素无先后顺序。

In [30]:
```
#可以对序列进行"切片"操作
mySeq1[1:3],mySeq2[1:3],mySeq3[1:3]
```

 序列有一些共同的特征和操作，参见本书【17 序列（P112/In[1]）】。

Out [30]: ('at', [2, 3], (12, 13))

In [31]:
```
#乘法操作在"序列"中有特殊含义
mySeq1*3
```

 序列的乘法有特殊含义——重复运算。

 关于序列的更多操作方法，参见本书【17 序列（P115/In[16]）】。

Out [31]: 'Data ScienceData ScienceData Science'

Python数据类型小结

	关键字	标志性符号	是否可变（允许局部替换）	是否为序列（支持序列操作）	强制类型转换函数
整型	int	无	否	否	int()
浮点型	float	小数点	否	否	float()
复数类型	complex	+/j	否	否	complex()
布尔类型	bool	True/False	否	否	bool()
字符（串）	str	单引号'或双引号"	否	是	str()
列表	list	方括号[ ]	是	是	list()
元组	tuple	圆括号( )和逗号,	否	是	tuple()
集合	set	花括号{ }	是	否	set()
不可变集合	frozenset	花括号{ }	否	否	frozenset()
字典	dict	花括号{ }和key	是	否	dict()

扩展 The Zen of Python（Python 禅）。

Python 编程的基本指导思想——The Zen of Python（Python 禅）的查看方法是在 Jupyter Notebook 中输入命令：import this。

5 变量

 常见疑问及解答。

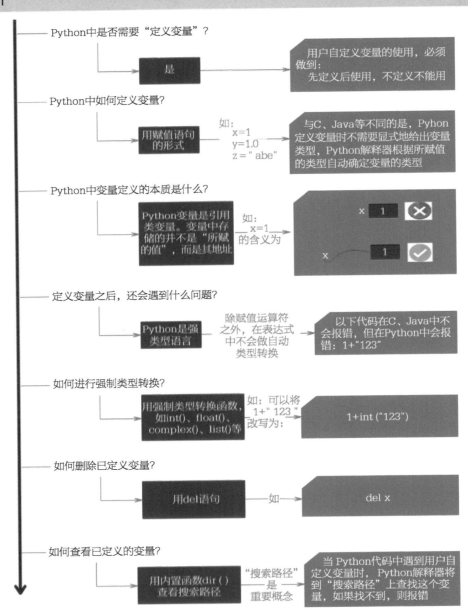

5.1 变量的定义方法

In [1]:

> **注意** 与 C、Java 等不同的是，Python 中用"赋值语句"的形式定义一个变量，即"用户不需要显式声明变量的数据类型"。

```
testBool = True
testInt = 20
testFloat = 10.6
testStr = "MyStr"
testBool, testInt, testFloat, testStr
```

Out [1]: (True, 20, 10.6, 'MyStr')

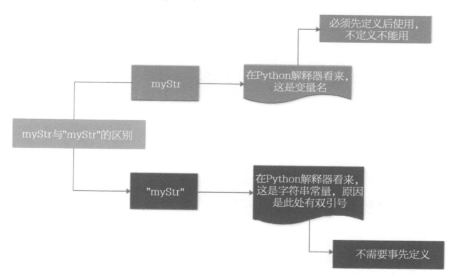

5.2 Python 是动态类型语言

In [2]:

> **注意** 以下代码在 C 或 Java 中会报错，但在 Python 中不会。

```
x = 10
x = "testMe"
```

5 变量

 Python 中变量不需要事先声明其所属类型，同一变量可以被赋值为不同的对象类型。

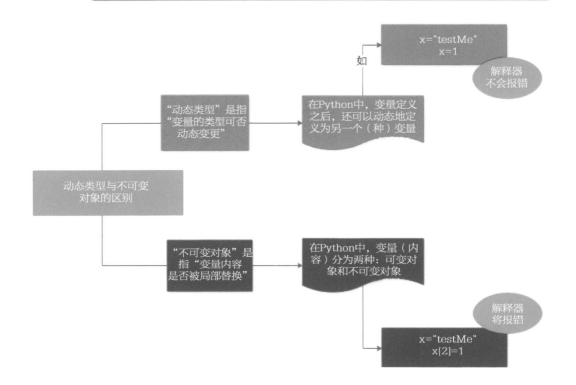

5.3 Python 是强类型语言

In [3]:

 Python 在"运算过程中不会自动进行数据类型转换"，除了 int、float、bool 和 complex 类型之间。

```
"3"+2    #出错，报错信息为TypeError
```

Out [3]:

```
TypeError                                 Traceback (most recent call last)
<ipython-input-3-0417f8c502b5> in <module>()
----> 1 "3"+2 #出错，报错信息为 TypeError

TypeError: must be str, not int
```

In [4]: 3+True #不报错

Out [4]: 4

In [5]: 3+3.3 #不报错

Out [5]: 6.3

In [6]: 3+(1+3j) #不报错

Out [6]: (4+3j)

5.4 Python 中的变量名是引用

In [7]:
> 提示：变量名代表的（或本质）是"值的一个引用"，而不是"变量的取值"。

```
i = 20
i = "myStr"
i = 30.1
i
```

Out [7]: 30.1

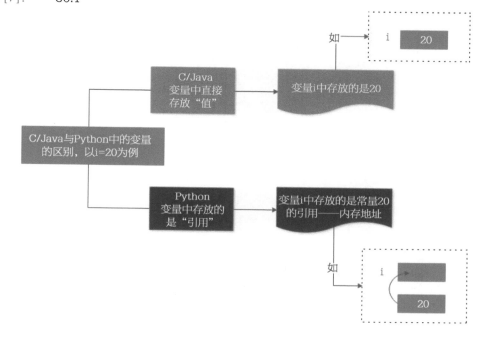

5.5 Python 中区分大小写

In [8]:
```
i = 20
I
```

 提示 | 定义的变量名为小写"i",而输出的变量名为大写"I"。

Out [8]:
```
NameError                                 Traceback (most recent call last)
<ipython-input-8-779b88bd1cc1> in <module>()
      1 i=20
----> 2 I

NameError: name 'I' is not defined
```

5.6 变量命名规范

In [9]:

 注意 | Python 变量命名要求如下。

原则 1:变量名只能包含字母、数字和下画线。
原则 2:变量名可以以字母或下画线开头,但不能以数字开头。
原则 3:不能用 Python 关键字作为变量名。

In [10]:
```
myvariable_2 = 0
```

In [11]:
```
2_myvariable = 0
```

 提示 | 报错原因:变量名以数字开头。

Out [11]:   File "<ipython-input-11-1eddbc5616ed>", line 1
    2_myvariable = 0
     ^
SyntaxError: invalid token

In [12]:

 如果用关键字作为变量名，会引起该关键字的含义发生改变，关键字的原有功能失效。

```
print = 0 #不报错
```

 从此，在当前 Jupyter Notebook kernel 会话范围内，print 代表的是 0，而不是输出功能/函数。

```
x = 0
print(x) #报错
```

 解决方法：重新启动"会话"。具体操作为：在 Jupyter Notebook 的菜单栏中依次选择 Kernel→Restart。

Out [12]:
TypeError   Traceback (most recent call last)
<ipython-input-12-713cdf74ab27> in <module>()
      3 x = 0
----> 4 print(x) #报错

TypeError: 'int' object is not callable

5.7 iPython 的特殊变量

In [13]:
```
x = 12+13
x
```

Out [13]:   25

In [14]:
```
# In[]变量
In[13]
```

 其实，In[]和 Out[]并非 Python 的变量，而是 iPython 为方便编辑代码和跟踪执行过程而给出的特殊变量。iPython 是 Python 的一种较为简单的编辑器和解释器。

5 变量

```
Out [14]:    'x = 12+13\nx'
```

In [15]:
```
# Out[]变量
Out[13]
```

```
Out [15]:    25
```

In [16]:
```
# 临时变量：_
_
#符号"_"代表的是"最近一个Out变量"。
```

```
Out [16]:    25
```

5.8 查看 Python 关键字的方法

In [17]:

 思路 用模块 keyword 中提供的属性——kwlist。

```
import keyword
keyword.kwlist
```

```
Out [17]:    ['False',
             'None',
             'True',
             'and',
             'as',
             'assert',
             'break',
             'class',
             'continue',
             'def',
             'del',
             'elif',
             'else',
             'except',
             'finally',
             'for',
```

```
 'from',
 'global',
 'if',
 'import',
 'in',
 'is',
 'lambda',
 'nonlocal',
 'not',
 'or',
 'pass',
 'raise',
 'return',
 'try',
 'while',
 'with',
 'yield']
```

5.9 查看已定义的所有变量

In [18]:

 可用 Python 内置函数 dir() 来显示搜索路径，即已确定的所有变量清单。

```
dir()
```

Out [18]:
```
['In',
 'Out',
 '_',
 '_1',
 '_13',
 '_14',
 '_15',
 '_16',
 '_17',
 '_4',
```

5 变量

```
'_5',
'_6',
'_7',
'__',
'___',
'__builtin__',
'__builtins__',
'__doc__',
'__loader__',
'__name__',
'__package__',
'__spec__',
'_dh',
'_i',
'_i1',
'_i10',
'_i11',
'_i12',
'_i13',
'_i14',
'_i15',
'_i16',
'_i17',
'_i18',
'_i2',
'_i3',
'_i4',
'_i5',
'_i6',
'_i7',
'_i8',
'_i9',
'_ih',
'_ii',
'_iii',
'_oh',
'exit',
'get_ipython',
```

```
 'i',
 'keyword',
 'myvariable_2',
 'print',
 'quit',
 'testBool',
 'testFloat',
 'testInt',
 'testStr',
 'x']
```

5.10 删除变量

In [19]:
```
i = 20
print(i)
```

 注意 此处需要重启 Jupyter Notebook 的 Kernel，否则报错，原因在于 In[12]中为"print=0"，即将 print 重定义为了变量名。

 技巧 重启 Kernel 的方法：在 Jupyter Notebook/Lab 菜单中依次选择 Kernel→Restart。

```
del i
```

 注意 在 Python 早期版本中，del 是语句（Statement），而不是函数（Function）。因此，写成 del(i)时会报错。

In [20]:
```
i
```

 提示 报错信息为 NameError，报错原因：在 In[1]中用 del i 语句删除了变量 i。

 注意 在数据分析和数据科学项目中，需要特别重视变量的当前值，当前值不一定与初始值一样。

5 变量

```
Out [20]:
NameError                                 Traceback (most recent call last)
<ipython-input-2-a7d1b7ed6c3a> in <module>()
----> 1 i

NameError: name 'i' is not defined
```

 提示 Python 之父 Guido 推荐命名规范包括如下几点。

1. 模块名和包名使用小写字母，并且以下画线分隔单词，如：regex_syntax, py_compile, _winreg
2. 类名或异常名采用每个单词首字母大写的方式，如：BaseServer, ForkingMixIn, KeyboardInterrupt
3. 全局或者类常量，全部使用大写字母，并且以下画线分隔单词，如：MAX_LOAD
4. 其余对象的命名，包括方法名、函数名、普通变量名，则采用全部小写字母，并且以下画线分隔单词的形式命名，如：my_thread
5. 若以上对象为私有类型，则使用下画线开头命名，如：__init__, __new__

 注意 考虑到数据分析和数据科学项目的特殊性，本书中的命名方法对 Guido 推荐命名规范进行了微调。

 思路 如何写出优秀的代码？PEP（Python Enhancement Proposal, Python 增强建议）是关于如何更好地完成 Python 编程的系列文档，其中 PEP8 为《Python 代码的编写规范（Style Guide for Python Code）》，PEP20 为《Python 禅（The Zen of Python）》，建议到官网 https://www.python.org/dev/peps/ 阅读相关资料。除了 PEP 系列文档，Google Style Guide 也是数据分析和数据科学实践中常用的代码编写规范。

6 语句书写规范

 常见疑问及解答。

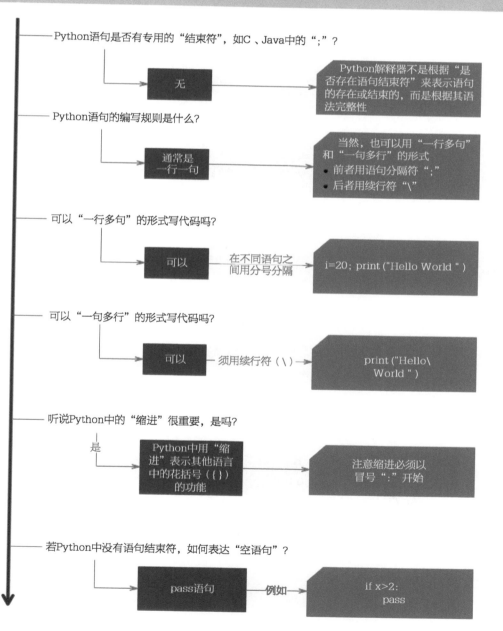

6 语句书写规范

 关于 Python 代码的编写规范,参见《PEP8-Style Guide for Python Code》以及《Google Python Style Guide》。

6.1 一行一句

In [1]:
```
i = 20
j = 30
k = 40
```

 与 C、Java 不同的是,Python 没有语句结束符,如";"。

6.2 一行多句

In [2]: 不同语句之间用";"分开。
```
i = 20;j = 30;k = 40
```

In [3]:
```
i;j;k
```
 输出结果与"i,j,k"不同。

Out [3]: 40

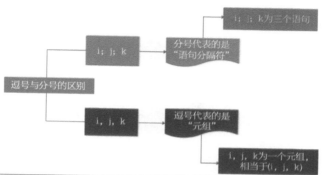

In [4]: i,j,k

> 注意　";" 与 "," 在 Python 中的区别：前者表示 "一行多句"，后者表示 "元组"，详见本书【15 元组】。

Out [4]: (20, 30, 40)

In [5]:
> 注意　初学者容易混淆分号和逗号的使用，如以下代码会报错。

print(i;j;k)

Out [5]:
```
  File "<ipython-input-5-326159b7de0d>", line 2
    print(i;j;k) #报错，SyntaxError: invalid syntax
            ^
SyntaxError: invalid syntax
```

> 提示　print(i;j;k) #报错，SyntaxError: invalid syntax，原因为分号在 Python 中表示的是语句分隔符。

6.3 一句多行

In [6]:
> 思路　续行符（\）。PEP8 建议，Python 代码中一行不应超过 79 个字符，否则应采用 "续行符" 写成多行。

print("nin \
hao")

Out [6]: nin hao

6.4 复合语句

In [7]:

注意　Python 中通过缩进方式（Indentation）表示 Java、C 中的{}。
Python 的缩进需要注意两点：

```
#第一，缩进开始之处必须加 ":";
#第二，代码的缩进与对齐方式很重要，同一个层级的代码的缩进方式
 应一致，即需要"对齐"。
sum = 0
for i in range(1,10):
    sum = sum+i
    print(i)
print(sum)
```

提示　PEP8 建议用 4 个空格（4 个半字空格）为一个缩进单位。

Out [7]:
1
2
3
4
5
6
7
8
9
45

In [8]:

提示　"缩进"在 Python 中的重要性：相当于 C、Java 中的{ }，同一层级代码的"缩进单位"必须相同，否则会报 IndentationError 错误信息。

```
a = 10
if a > 5:
    print("a+1 = ",a+1)
print("a = ",a)
```

Out [8]:   a+1 = 11
           a = 10

  Python 中没有固定的"缩进单位",但是通常以 4 个空格为一个缩进单位。

  凡是缩进开始之处,必须有冒号(:)。

6.5 空语句

In [9]:
```
x = 1
y = 2
if x>y:
    pass
else:
    print(y)
```

Out [9]:   2

  通常,Python 是"可执行的伪代码",空语句需要用 pass 语句来"占位",否则报错。

  与 C、Java 不同的是,Python 中的空语句用 pass 语句表示,并非用";"表示。

  Python 的两个重要版本。

目前,Python 有两个重要版本:Python 2 和 Python 3。二者在语法上有很多不同,本书遵循的是 Python 3 的语法。二者之间的主要区别及自动转换方法参见 https://docs.python.org/3/howto/pyporting.html。

7 赋值语句

常见疑问及解答。

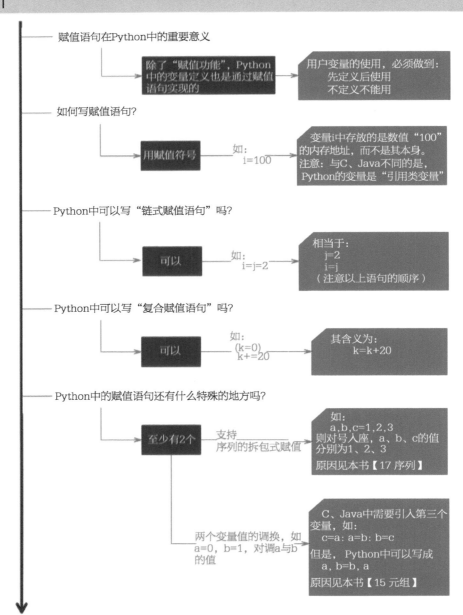

7.1 赋值语句在 Python 中的重要地位

In [1]:

提示 除了赋值功能，赋值语句还用于"定义一个新变量"，如

```
i = 1
i
```

Out [1]:   1

7.2 链式赋值语句

In [2]:
```
i = j = 2
```

注意 赋值运算符的结合方向为"从右到左"。
```
i
j
```

Out [2]:   2

In [3]:

提示 链式赋值语句"i = j = 2"在功能上等价于如下代码。
```
j = 2
i = j
i
j
```

Out [3]:   2

7 赋值语句

7.3 复合赋值语句

In [4]:
 提示 | Python 语言中的复合赋值语句的写法与 C、Java 类似。

```
i = 1
i += 20
i
```

Out [4]: 21

In [5]:

x *= 20 + 5 的计算过程

第一步：对赋值符号右侧部分整体加圆括号

x *= (20 + 5)

第二步：将等号左侧移动到新增( )的左侧，生成一个"新表达式"

x * (20 + 5)

第三步：将第二步生成的"新表达式"的结果赋值给最初变量x

x = x * (20 + 5)

7.4 序列的拆包式赋值

In [6]:

 思路 | 拆包式赋值的赋值规则为"对号入座"。

 提示 | 关于序列及其拆包式赋值的更多内容，参见本书【17 序列（P114/In[13]）】。

```
a,b,c=1,2,3
a,b,c
```

 此处输出结果为元组,即带圆括号,参见本书【15 元组（P95/In[1]）】。

Out [6]: (1, 2, 3)

7.5 两个变量值的调换

In [7]:

 与 C、Java 不同的是,Python 中可以直接"对调"两个变量的值 a,b=b,a。原因如下:a,b 相当于(a,b),是一个元组,但(b,a)是另一个元组,即 a,b=b,a 相当于(a,b)=(b,a),属于元组的拆包式赋值,详见【15.3 基本用法】。

```
a=1
b=2
a,b=b,a
a,b
```

 在 C、Java 中,两个变量 a 和 b 值的对调操作需要引入第三个变量(c),如:

```
c = a;a = b;b = c;
```

Out [7]: (2, 1)

 教学互动。

学生：请问老师，Python 中可以对同一个变量先后赋值两个不同类型的数值吗？
老师：这种问题你自己试着写代码就知道了。

8 注释语句

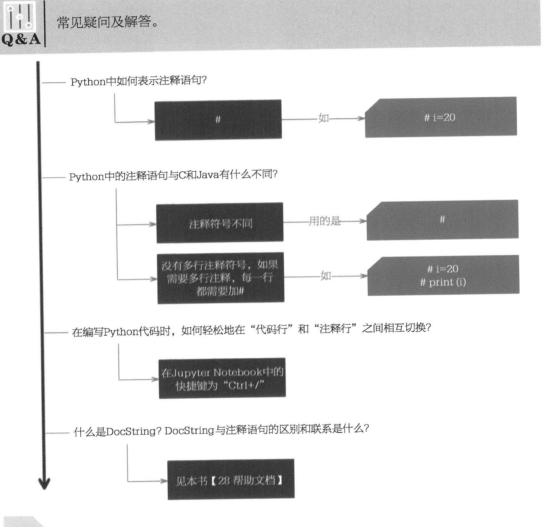

8.1 注释方法

In [1]: 注意：与 C 和 Java 不同的是，Python 中的注释符号为 "#"，而且无 "多行注释符号"。

i = 20

> 👁 **提示** Python 解释器将跳过所有的以 "#" 开始的注释行代码。

In [2]:
```
# i=20
i
```

> 👁 **提示** 解释器报错原因在于，该变量的定义部分（即 i=20）是注释行。

Out [2]:
```
NameError                                 Traceback (most recent call last)
<ipython-input-2-f7d09bb3600b> in <module>()
      1 # i=20
      2
----> 3 i #提示：解释器报错原因在于，该变量的定义部分是注释行。

NameError: name 'i' is not defined
```

8.2 注意事项

In [3]:

> 👁 **提示** Python 中没有多行注释符号，如果需要多行注释，则每一行需要加注释符号 "#"。

```
# i = 20
# i
```

> 🔧 **技巧** 在 Jupyter Notebook 中，可以用快捷键 Ctrl+/ 在 "注释行" 和 "代码行" 之间轻松切换。注意，在中文输入法状态下，无法使用 Jupyter Notebook 的快捷键。

> 💡 **注意** 关于 docString，参见本书【28 帮助文档（P181/In[2]）】。在 Python 中，所有标点符号必须为英文标点符号。在 Jupyter Notebook 中，所有快捷键应在英文输入状态下使用。

9 运算符

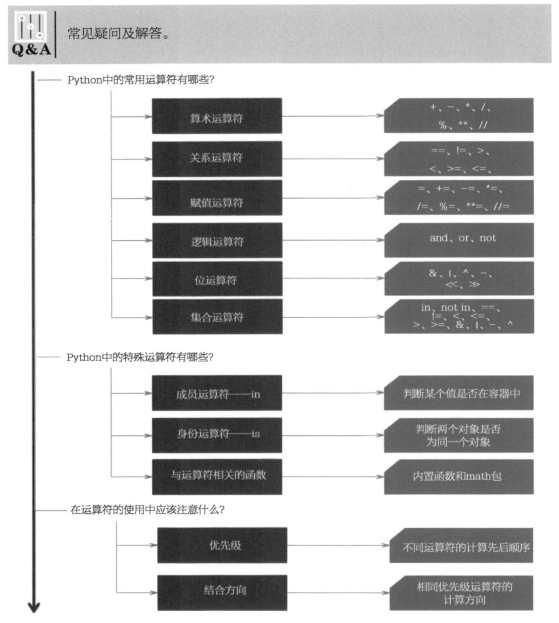

算术运算符、关系运算符、赋值运算符、逻辑运算符、位运算符、集合运算符分别如表 9-1 至表 9-6 所示。

表 9-1 算术运算符（x=2, y=5）

运算符	含义	实例	结果
+	加	x + y	7
-	减	x - y	-3
*	乘	x * y	10
/	除	y / x	2.5
%	取余	y % x	1
//	整除	y//2	2
**	幂	x**y	32

表 9-2 关系运算符（x=2, y=5）

运算符	含义	实例	计算结果
==	等于	x == y	False
!=	不等于	x != y	True
>	大于	x > y	False
<	小于	x < y	True
>=	大于等于	x >= y	False
<=	小于等于	x <= y	True

表 9-3 赋值运算符

运算符	实例	等价
=	y=x	y=x
+=	y+=x	y=y+x
-=	y-=x	y=y-x
*=	y*=x	y=y*x
/=	y/=x	y=y/x
%=	y%=x	y=y%x
**=	y**=x	y=y**x
//=	y//=x	y=y//x

表 9-4 逻辑运算符（x=2, y=5）

运算符	含义	实例	结果
and	与	x and y	5
or	或	x or y	2
not	非	not (x and y)	False

9 运算符

表 9-5 位运算符（x=2, y=5；注：查看对应二进制的方法，用内置函数 bin()）

运算符	含义	实例	结果
&	按位与	x & y	0
|	按位或	x | y	7
^	按位异或	x ^ y	7
~	按位取反	~x	−3
<<	左移	x << y	64
>>	右移	x >> y	0

表 9-6 集合运算符

数学符号	Python符号	说明
∈	in	是……的成员
∉	not in	不是……的成员
=	==	等于
≠	!=	不等于
⊂	<	是……的（严格）子集
⊆	<=	是……的子集（包括非严格子集）
⊃	>	是……的（严格）超集
⊇	>=	是……的超集（包括非严格超集）
∩	&	交集
∪	|	合集
−或\	−	差补或相对补集
△	^	对称差分

Python运算符的优先级如图9-1所示。

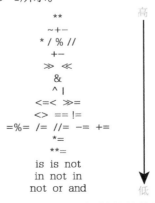

```
            **
          ~ + −
         * / % //
           + −
           ≫ ≪
            &
           ^ |
        <= < ≫ >=
         <> == !=
   =%= /= //= −= +=
           *=
          **=
        is  is not
        in  not in
        not or and
```

图 9-1 Python 运算符的优先级

9.1 特殊运算符

In [1]:
```
#除法
x = 2
y = 5
y / x
```

> 提示：PEP8 建议运算符前、后各空一格。

Out [1]: 2.5

In [2]:
```
#取余
x = 2
y = 5
y % x
```

Out [2]: 1

In [3]:
```
#整除
x = 2
y = 5
y // x
```

Out [3]: 2

In [4]:
```
#幂次
x = 2
y = 5
x ** y
```

Out [4]: 32

In [5]:
```
#相等
x = 2
y = 5
x == y
```

Out [5]: False

7 ÷ 2 = 3 1
相当于 7//2　　相当于 7%2

9 运算符

In [6]:
```
#不等
x = 2
y = 5
x != y
```

Out [6]: True

In [7]:
 Python 中有特殊运算符：is、is not、in、not in。

 is 运算符的功能：判断是否指向同一个引用。

```
x = 2
y = 5
x is y
```

Out [7]: False

In [8]:
```
x = 2
y = 5
x is not y
```

Out [8]: True

In [9]:
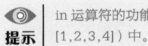 in 运算符的功能：判断某个变量（如 x）是否在给定容器（如 [1,2,3,4]）中。

```
x in [1,2,3,4]
```

Out [9]: True

In [10]:
```
y in [1,2,3,4]
```

Out [10]: False

In [11]:
```
x not in [1,2,3,4]
```

Out [11]: False

In [12]:
```
#复合赋值运算符
x = 2
y = 5
y //= x
```

> **提示** 相当于 y = y // x。

```
print(x,y)
```

Out [12]: 2 2

In [13]:

> 💡 **注意** 以下计算中，y 的结果为 0，而不是 10，原因见本书【7 赋值语句（P46/In[5]）】。

```
x = 2
y = 5
y //= x + 8
print(y)
```

Out [13]: 0

In [14]:
```
#逻辑运算符
x = True
y = False
x and y
```

Out [14]: False

In [15]:
```
x = True
y = False
x or y
```

Out [15]: True

In [16]:
```
x = True
not x
```

Out [16]: False

In [17]:
```
#位运算符
```

9 运算符

> **注意** 位运算符和逻辑运算符是两个不同概念。

```
x = 2
y = 3
print(x,y)
print(bin(x),bin(y))
```

> **提示** 可以用内置函数 bin() 将十进制数据转换为二进制数据。

Out [17]: 2 3
0b10 0b11

In [18]:
```
x = 2
y = 3
x&y
```

> **提示** &是位运算符，含义为"按位与"计算。

Out [18]: 2

In [19]:
```
x = 2
y = 3
bin(x&y)
```

Out [19]: '0b10'

In [20]:
```
x = 2
y = 3
bin(x|y)
```

Out [20]: '0b11'

In [21]: `bin(x^y)`

Out [21]: '0b1'

In [22]: `bin(~x)`

Out [22]: `'-0b11'`

In [23]:
```
x = 2
y = 3
bin(x<<y)
```

Out [23]: `'0b10000'`

In [24]:
```
x = 2
y = 3
bin(x>>y)
```

Out [24]: `'0b0'`

9.2 内置函数

In [25]:

 提示 内置函数（Built-In Function，BIF）是指内置在 Python 解释器中的函数，可以直接通过函数名调用它。

`pow(2,10)`

 注意 pow()为内置函数，但 sin()不是内置函数。

 技巧 查看内置函数的方法：
`dir(__builtins__)`

Out [25]: 1024

In [26]:
`round(2.991)`

 提示 四舍五入函数：round(number, ndigits)。其功能为对其第一个参数 number 进行四舍五入，小数点后边保留 ndigits 个有效位，ndigits 的默认值是 0。

Out [26]: 3

In [27]:
```
round(2.991,2)
```

 提示 | 参数"2"的含义为"小数点后保留2个有效位"。

 思路 | 可以通过"?round"或"round?"查看函数round()的帮助信息,系统给出的帮助信息如下:

round(number[,ndigits])

放在[]中的形式参数为可选参数,如ndigits

Out [27]: 2.99

9.3 math 模块

In [28]:

 提示 | 在 Python 中,很多常用数学函数(如 sin()、cos()等)并非内置函数,而是放在 math 模块中。

```
import math
math.sin(5/2)
```

Out [28]: 0.5984721441039564

In [29]:
```
import math
math.pi
```

 提示 | 显示π的方法。

Out [29]: 3.141592653589793

In [30]:
```
import math
math.sqrt(2.0)     #求平方根
```

Out [30]: 1.4142135623730951

In [31]:
```
import math
math.sqrt(-1)
```
 在 math 模块中求负数的平方根时会报错。

Out [31]:
```
ValueError                                Traceback (most recent call last)
<ipython-input-31-4e02a0210c6b> in <module>()
      1 import math
----> 2 math.sqrt(-1)

ValueError: math domain error
```

In [32]:
```
import cmath
cmath.sqrt(-2)
```
 复数对应的函数在另一个模块 cmath 中。

Out [32]: 1.4142135623730951j

9.4 优先级与结合方向

In [33]:
 不同运算符的优先级不同，结合方向也可能不同。
```
2**2**3
```
 在 Python 中，"2**2**3" 与 "(2**2)**3" 不同

Out [33]: 256

In [34]:
```
(2**2)**3
```

Out [34]: 64

9 运算符

In [35]:
```
x=2+3
x
```

> **注意** 赋值运算符的优先级和结合方向。

Out [35]: 5

In [36]:
```
1+2 and 3+4
```

> **提示** 请分析表达式"1+2 and 3+4"的结果为"7"的原因。

Out [36]: 7

> **扩展** 开放学习Python。

Python 开发人员，包括数据分析师和数据科学家，应积极加入领域重要的专业社区、邮件列表和开源项目。不要自己一个人孤立地"闭门式学习"，应打开自己的世界，与同行一起共同学习。例如，Stack Overflow 是数据科学家常用的社区，包括各种疑难问题的专业级解答和讨论，Stack Overflow 界面如图 9-2 所示。

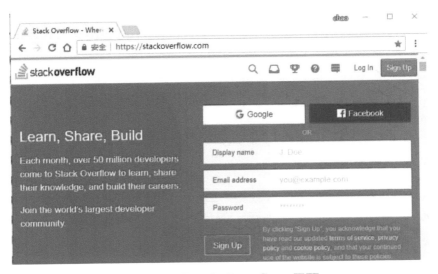

图 9-2 Stack Overflow 界面

10 if 语句

常见疑问及解答。

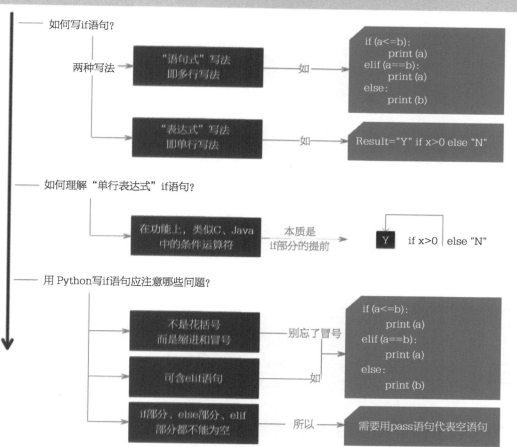

10.1 基本语法

In [1]:

在 Python 中，用"缩进"来代替 C 和 Java 中的"{}"，即复合语句的功能。

```
a = 2
b = 3
if(a < b):
```

10 if 语句

```
    print("a 小于 b")
else:
    print("a不小于b")
```

 注意　Python 中，每次缩进的开始之处必须有 ":"，即缩进必须与 ":" 一起用。

In [2]:

提示　与 C 和 Java 类似，Python 也支持 if 语句的嵌套。

```
if(a <= b):
    if(a < b):
        print(a)
    else:
        print(a)
else:
    print(b)
```

Out [2]:　2

10.2 elif 语句

 提示　与 C 和 Java 不同的是，Python 中的 if 语句可以带有 elif 部分。更有趣的是，try- catch、while、for 等语句中均可以出现 else 部分。在 Python 中，else 语句代表的是以正常方式（不是 break，continue 或抛出异常等）退出语句。

In [3]:

 注意　"elif" 不能写成 elseif。

```
if(a <= b):
    print(a)
elif(a == b):
    print(a)
else:
    print(b)
```

Out [3]:　2

10.3 if 与三元运算

In [4]:

> **思路** if 可以写成单行表达式,相当于 C 和 Java 中的三元条件运算符 "?:"。

```
x = 0
Result = "Y" if x > 0 else "N"
```

> **注意** 在三元运算中,Y 的位置提前至 if 语句之前。

```
Result
```

> **思路** 在 Python 中,if 语句、for 语句和函数均可以写成单行,分别称为三元运算符、列表推导式和 lambda 函数。

Out [4]: 'N'

In [5]:
```
x = 1
Result = "Y" if x > 0 else "N"
Result
```

Out [5]: 'Y'

Python中的单行写法之"if语句"

原始if语句——多行if语句	三元运算——单行if语句

原理:

```
if 【1】:
    变量名 = 【2】
else:
    变量名 = 【3】
```

→ 变量名 = 【2】 if 【1】 else 【3】

【提示】if部分移至最前面

实例:

```
if x>0:
    Result = "Y"
else:
    Result = "N"
```

→ Result="Y" if x>0 else "N"

10 if 语句

10.4 注意事项

In [6]:

注意 if 语句中的每部分均不能为空,否则 Python 解释器将报错,因为"Python 是可执行的伪代码",参见本书【13 pass 语句(P74/In[1])】。

```
if(a <= b):
```

提示 能否写一个空行呢?不能。
IndentationError: expected an indented block

```
else:
    print(b)
```

提示 pass 语句相当于其他语言中的空语句。

Out [6]:
```
  File "<ipython-input-6-84ea399e22ec>", line 4
    else:
       ^
IndentationError:expected an indented block
```

In [7]:

提示 纠正方法:用空语句 pass。

```
if(a <= b):
    pass      #无报错
else:
    print(b)
```

In [8]:

提示 如何用 Python 语句判断是否为"闰年"呢?建议如下。

```
import calendar
calendar.isleap(2019)
```

思路与技巧 软件开发项目与数据分析项目具有本质的区别。因此，在数据分析与数据科学项目中，不能将 Python 当作 C 或 Java 来用。以"判断是否闰年"为例，不要试图用 Python 来翻译 Java 或 C 的代码。

Out [8]:　　False

扩展 数据分析师和数据科学家常用的开发工具。

　　数据分析师和数据科学家常用的开发工具有两种：一种是开源语言工具，如 Python 和 R 等；另一种是商业工具，如 IBM Cognitive Class Lab 等。Gartner 每年定期出版专题研究报告 *Magic Quadrant for Data Science and Machine-Learning Platforms* 对数据科学与大数据分析中常用商业工具进行动态排名与评估分析，建议读者阅读和关注该系列报告。

　　IBM Cognitive Class Lab（原名 IBM Data Scientist Workbench）以"集成"的方式提供了 Python 大数据分析和数据科学实践所需的计算平台、数据、开发环境、常用工具、典型案例及学习资源，其界面如图 10-1 所示。

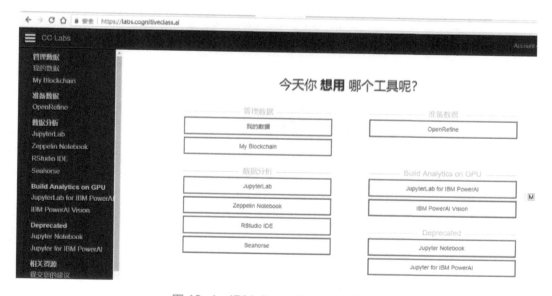

图 10-1　IBM Cognitive Lab 界面

11 for 语句

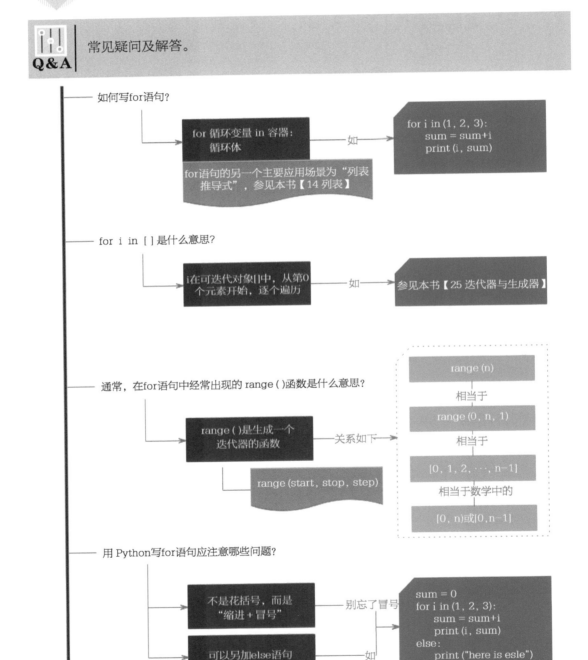

11.1 基本语法

In [1]:

 与 C、Java 不同的是,Python 的 for 语句的写法只有一种:【for… in…】。

```
sum = 0
for i in (1,2,3):
    sum = sum+i
    print(i,sum)
```

 别忘了冒号,注意对齐方式,参见本书【6 语句书写规范(P42/In[7])】。

 in 后面为"可迭代对象或迭代器"。在此,()代表的是元组,元组是可迭代对象,参见本书【25 迭代器与生成器(P165)】。

 for 语句之前,需要对 sum 赋值,否则报错未定义变量。

Out [1]:　1 1
　　　　　2 3
　　　　　3 6

11.2 range()函数

In [2]:

 range()函数经常出现在for语句中的in之后,如range(1,10)。

 range()函数的返回值为一个"range迭代器"。

 关于迭代器见本书【25 迭代器与生成器(P165)】。

```
range(1, 10)
```

range(n)
相当于
range(0, n, 1)
相当于
[0, 1, 2, …, n−1]
相当于数学中的
[0,n),即左包含右不包含的结构

11 for 语句

Out [2]:   range(1, 10)

In [3]:

 技巧 为了显示迭代器的内容，可以用 list() 函数将 "range 迭代器" 转换为列表。

```
myList = list(range(1,10))
myList
```

注意 range(1,10) 函数的返回值中有 "1"，但无 "10"，参见本书 【14 列表（P76）】。

Out [3]:   [1, 2, 3, 4, 5, 6, 7, 8, 9]

11.3 注意事项

In [4]:

 提示 与 C 和 Java 不同的是，Python 的 for 语句可以带 else 部分。

```
sum = 0
for i in (1,2,3):
    sum = sum+i
    print(i,sum)
else:
    print("here is the else statement")
```

Out [4]:   1 1
    2 3
    3 6
    here is the else statement

In [5]:

 提示 与 C、Java 等类似，Python 的 for 语句支持 break 和 continue 语句。

```
myList = list(range(1,10))
for j in [1,3,4,5]:
    print(myList[j])
```

Out [5]: 2
4
5
6

In [6]:
> 注意 break 和 continue 的区别，前者为"跳出循环体"，后者为"跳在循环体之内"，如下面的箭头所示。

```
for k in range(0, 16, 2):
    if (k == 8):
        break
    print (k)
```

Out [6]: 0
2
4
6

In [7]:
> 提示 与 break 不同的是，continue 的含义为"跳在循环体之内"，即跳过循环中的当前一步，如下面的箭头所示。

```
for k in range(0, 16, 2):
    if (k == 8):
        continue
    print (k)
```

Out [7]: 0
2
4
6
10
12
14

11 for 语句

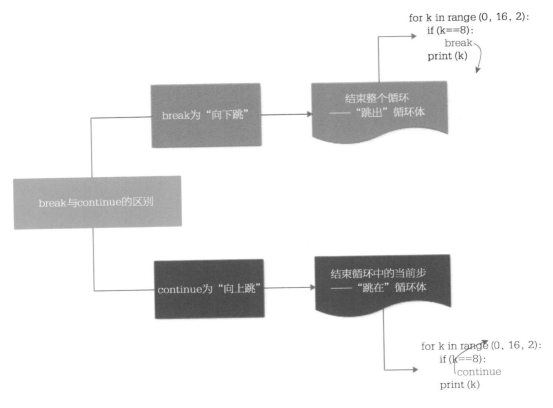

 Python 中可以带有 else 子句的地方。

与 C 和 Java 不同的是，Python 中有如下几个地方可以带有 else 子句。
（1）在 if 语句中，见本书【10 if 语句】。
（2）在 for 语句中，见本书【11 for 语句】。
（3）在 while 语句中，见本书【12 while 语句】。
（4）在 try-catch 语句中，见本书【29 异常与错误】。

12 while 语句

  常见疑问及解答。

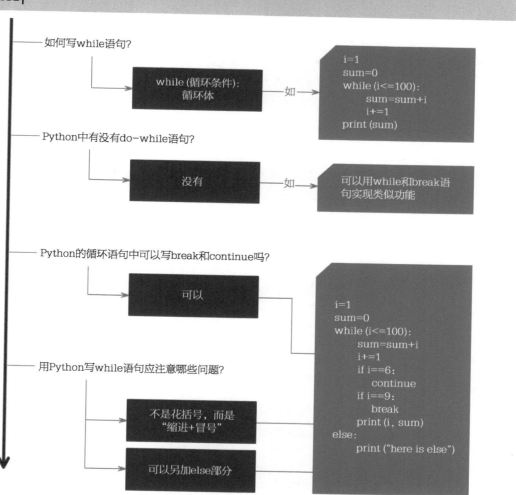

12.1 基本语法

In [1]:   在 Python 中，while 语句的写法单一，且没有 do-while 语句。

12 while 语句

```
i = 1
sum = 0
while(i <= 100):
    sum = sum+i
    i += 1
print(sum)
```

> **注意**：别忘了"冒号+缩进"。

Out [1]: 5050

12.2 注意事项

In [2]:
```
i = 1
sum = 0
while(i <= 10):
    sum = sum+i
    i += 1
    if i == 6:
        continue
    if i == 9:
        break
    print(i,sum)
else:
    print("here is else")
```

> **注意**：break 和 continue 语句的区别：前者为"跳出"循环体，后者为"跳在"循环体。如上面箭头指向所示。

Out [2]: 2 1
3 3
4 6
5 10

```
                    7 21
                    8 28
In [3]:     i = 1
            sum = 0
            while(i <= 10):
                sum = sum+i
                i += 1
                print(i,sum)
            else:
                print("here is else")
```

 与 C 和 Java 不同的是，Python 的 while 语句中可以带有 else 部分。while 语句的 else 部分只有在正常方式（而不是 break 或 continue 的方式）退出循环时才运行。

```
Out [3]:    2 1
            3 3
            4 6
            5 10
            6 15
            7 21
            8 28
            9 36
            10 45
            11 55
            here is else
```

 Python 代码类型的分辨方法。

（1）以 "#" 开始的代码是注释语句，见本书【8 注释语句】。

（2）以 "%" 开始的代码是魔术命令（并非为 Python 语法，属于 iPython、Jupyter Notebook 语法），见本书【32 魔术命令】。

（3）以 "@" 开始的代码是装饰器，见本书【31 面向对象编程】。

13 pass 语句

常见疑问及解答。

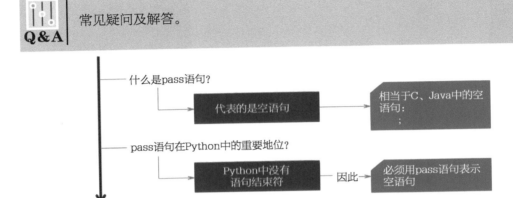

13.1 含义

In [1]:
 Python 中的空语句必须用 pass 语句表示，否则出错，原因："Python 是可执行的伪代码"。

```
a = 10
b = 11
if(a <= b):
```

 在此能否写一个空行呢？不能。
IndentationError: expected an indented block

```
else:
    print(b)
```

 Python 中的 pass 语句相当于 C 和 Java 的空语句 "；"。

Out [1]:   File "<ipython-input-1-6afc05c62f97>", line 6
      else:
         ^
    IndentationError: expected an indented block

13.2 作用

In [2]:

> **提示** In[1]中的错误的纠正方法：用 pass 语句表示空语句。在 Python 中，pass 语句的含义为空语句。

```
a = 10
b = 11
if(a <= b):
    pass
else:
    print(b)
```

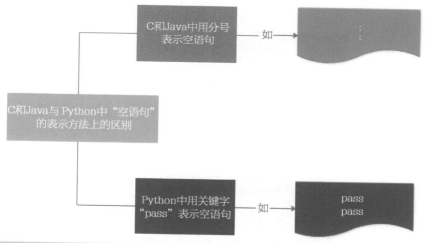

 Python 编程中常用的网站。

Python 编程中常用的网站主要如下。

- ❖ Python 官网：https://www.python.org/
- ❖ iPython 官网：https://ipython.org/
- ❖ Jupyter Notebook 官网：http://jupyter.org/
- ❖ Anaconda 官网：https://www.anaconda.com/
- ❖ PyPI（Python Package Index）官网：https://pypi.org/

14 列表

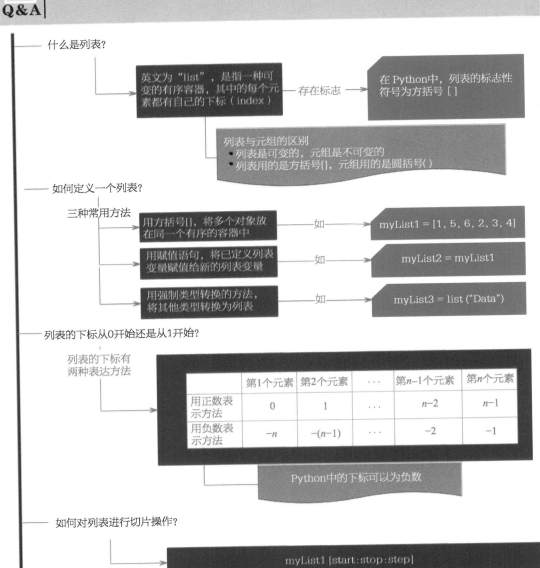

— 列表的常用运算有哪些?

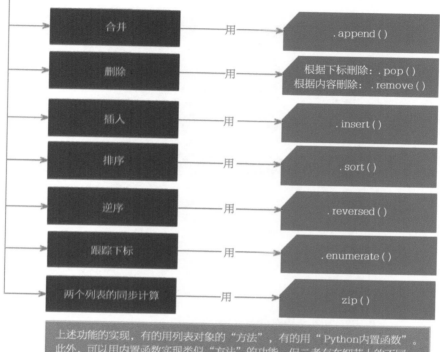

上述功能的实现,有的用列表对象的"方法",有的用"Python内置函数"。此外,可以用内置函数实现类似"方法"的功能,但二者存在细节上的不同。如列表的方法.sort()对应的内置函数有sorted(),虽然都可以用于列表的排序,但后者不改变列表本身,而生成另一个新的列表;前者则相反

— 什么是列表的推导式?

— Python中的列表编程应注意哪些问题?

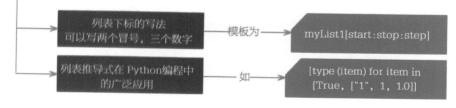

14 列表

14.1 定义方法

In [1]:
> 第一种方法：用[]。

```
myList1 = [21,22,23,24,25,26,27,28,29]
myList1
```
> 注意：在 Python 基础语法中，()、[]、{}分别代表元组、列表和集合/字典。

Out [1]: [21, 22, 23, 24, 25, 26, 27, 28, 29]

In [2]:
> 第二种方法：用赋值语句，即将已定义列表变量赋值给新的列表变量。

```
myList2 = myList1
myList2
```

Out [2]: [21, 22, 23, 24, 25, 26, 27, 28, 29]

In [3]:
> 第三种方法：用强制类型转换的方法，将其他类型的对象转换为"列表"。

```
myList3 = list("Data")
myList3
```

Out [3]: ['D', 'a', 't', 'a']

In [4]:
> 注意：在 Python 中可以用负下标/负索引。"正下标"和"负下标"的对应关系如下：

```
myList1[-1]
```

Out [4]: 29

In [5]:
```
myList1[-9]
```

Out [5]: 21

In [6]: myList1[9]

 报错原因是下标超出了边界。

 Python 中"正下标"与"负下标"的区别：正下标是从 0 开始的，从左到右编号；负下标从 -1 开始，从右到左编号。

Out [6]:

IndexError   Traceback (most recent call last)
<ipython-input-6-6a89397133c8> in <module>()
----> 1 myList1[9]

IndexError: list index out of range

myList1 = [21, 22, 23, 24, 25, 26, 27, 28, 29]

读取22的方法有两种：
- myList1[1]
- myList1[-8]

	第1个元素	第2个元素	...	第n-1个元素	第n个元素
用正数表示方法	0	1	...	n-2	n-1
用负数表示方法	-n	-(n-1)	...	-2	-1

即

myList1	21	22	23	24	25	26	27	28	29
正下标	0	1	2	3	4	5	6	7	8
负下标	-9	-8	-7	-6	-5	-4	-3	-2	-1

14.2 切片操作

In [7]: myList1

 查看 myList1 的当前值。
数据分析和数据科学项目中需要注意变量的当前值。

Out [7]: [21, 22, 23, 24, 25, 26, 27, 28, 29]

14 列表

In [8]: myList1[1:8]

> 提示：在 Python 中，列表的切片操作是通过下标进行的，"切片操作"的模式为 "start:stop:step"。
>
> 注意：Python 序列的下标中出现 ":" 时，多数情况为对其进行"切片操作"，参见本书【17 序列】。
>
> 思路：在"切片操作"中，start、stop、step 参数均可省略。

Out [8]: [22, 23, 24, 25, 26, 27, 28]

In [9]: myList1[1:8:2]

> 提示：此处，step 参数的值设置为 2。

Out [9]: [22, 24, 26, 28]

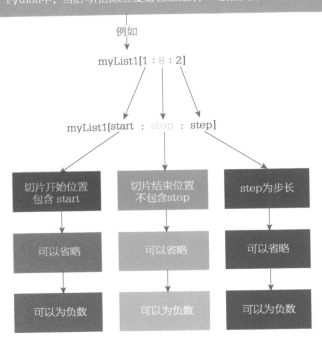

In [10]: myList1[:5]

 提示 | 省略了 start 和 step 参数。

 注意 | 在"切片操作"中不包括 stop 位置上的要素，如本例中不含下标为 5 的元素，即取值为 26 的元素。

Out [10]: [21, 22, 23, 24, 25]

In [11]: myList1[:]

 提示 | 省略了 start、stop、step 参数。

Out [11]: [21, 22, 23, 24, 25, 26, 27, 28, 29]

In [12]: myList1[2:]

 提示 | 省略了 stop 和 step 参数。

Out [12]: [23, 24, 25, 26, 27, 28, 29]

In [13]: myList1[:-1]

 提示 | "切片操作"支持负下标。

Out [13]: [21, 22, 23, 24, 25, 26, 27, 28]

14.3 反向遍历

In [14]: myList1

Out [14]: [21, 22, 23, 24, 25, 26, 27, 28, 29]

14 列表

In [15]: `myList1[::-1]`

 可以用下标[::-1]（即 step=-1）的方式实现列表的反向遍历。

 在此有两个冒号。

Out [15]: [29, 28, 27, 26, 25, 24, 23, 22, 21]

In [16]: `myList1`

 切片操作不会改变列表本身。

Out [16]: [21, 22, 23, 24, 25, 26, 27, 28, 29]

In [17]: `myList1[:-1]`

 [:-1]与[:n-1]的含义一样，在数据科学项目中经常出现下标为 -1 的情况，表示的是该下标的最大取值。

Out [17]: [21, 22, 23, 24, 25, 26, 27, 28]

In [18]: 实现列表的反向遍历，还可以用内置函数 reversed()或列表方法 reverse()。

`reversed(myList1)`

 reversed()函数的返回值为一个迭代器，可以用 list()强制转换的方式显示其取值内容。

 关于迭代器，参见本书【25 迭代器与生成器（P165）】。

Out [18]: `<list_reverseiterator at 0x1eba97dfac8>`

In [19]: `list(reversed(myList1))`

Out [19]: [29, 28, 27, 26, 25, 24, 23, 22, 21]

In [20]:
> 查看 myList1 的当前值。
>
> myList1

Out [20]: [21, 22, 23, 24, 25, 26, 27, 28, 29]

In [21]:
> 内置函数 reversed() 与列表方法 reverse() 的区别：前者不改变列表本身，后者将改变列表本身。
>
> myList1.reverse()
> myList1

Out [21]: [29, 28, 27, 26, 25, 24, 23, 22, 21]

14.4 类型转换

In [22]:
> list("chaolemen")
>
> 用 list() 进行强制类型转换。

Out [22]: ['c', 'h', 'a', 'o', 'l', 'e', 'm', 'e', 'n']

14.5 extend 与 append 的区别

In [23]:
> #【1】列表的 "+" 运算
> myList1 = [21,22,23,24,25,26,27,28,29]
> myList2 = myList1
> myList1 + myList2

Out [23]: [21, 22, 23, 24, 25, 26, 27, 28, 29, 21, 22, 23, 24, 25, 26, 27, 28, 29]

In [24]:
> myList1 = [21,22,23,24,25,26,27,28,29]
> myList2 = myList1
> myList1.extend(myList2)

14 列表

 列表的"+"运算相当于列表的 extend() 方法。

myList1

Out [24]: [21, 22, 23, 24, 25, 26, 27, 28, 29, 21, 22, 23, 24, 25, 26, 27, 28, 29]

In [25]:
#【2】列表的方法 append()
myList1.append(myList2)
myList1

 列表的方法 append() 与 extend() 的区别：前者"以成员身份追加"，而后者则"平等追加"。

Out [25]: [21, 22, 23, 24, 25, 26, 27, 28, 29, 21, 22, 23, 24, 25, 26, 27, 28, 29, [...]]

In [26]:
#【3】两个列表的并列相加计算
myList1 = [1,2,3,4,5,6,7,8,9]
myList3 = [11,12,13,14,15,16,17,18,19]
[i + j for i, j in zip(myList1, myList3)]

 zip() 函数的功能为并行迭代。

列表推导式必须放在[]中，参见本书【14.6 列表推导式（P84/In[27]）】。

Out [26]: [12, 14, 16, 18, 20, 22, 24, 26, 28]

14.6 列表推导式

In [27]:

与 C 和 Java 不同的是，Python 中有"列表推导式"的概念，可以简化复杂的 for 语句。

>
> **思路** 由于 Python 中有列表推导式、ufunc 函数、向量化计算等机制，因此基于 Python 的数据科学项目中一般不出现复杂的 for 语句。
>
>
> **提示** 可以采用"列表推导式"快速生成"列表"。
>
> ```
> [2 for i in range(20)]
>     #先做 range()，再做 i，最后做 2
>     #注意：（1）列表推导式必须放在[]中；（2）for 之前是一个表达式。
> ```

Out [27]: [2, 2, 2, 2, 2, 2, 2, 2, 2, 2, 2, 2, 2, 2, 2, 2, 2, 2, 2, 2]

In [28]: `[i for i in range(1, 21)]`

Out [28]: [1, 2, 3, 4, 5, 6, 7, 8, 9, 10, 11, 12, 13, 14, 15, 16, 17, 18, 19, 20]

In [29]: `[i for i in range(1, 21, 2)]`

Out [29]: [1, 3, 5, 7, 9, 11, 13, 15, 17, 19]

In [30]:
```
range(10)
```
>
> **提示** range(10)相当于 range(0,10)。

Out [30]: range(0, 10)

In [31]:
```
list(range(0,10,2))
```
>
> **提示** 此处，0、10、2 分别为迭代器的 start、stop 和 step 参数。

Out [31]: [0, 2, 4, 6, 8]

In [32]: `[type(item) for item in [True,"1",1,1.0]]`

Out [32]: [bool, str, int, float]

14 列表

Python中的单行写法之【列表推导式】

原始for语句——多行

原理：
```
for 【1】 in 【2】:
        【3】
```

列表推导式——单行for语句

```
[ 【3】 for 【1】 in 【2】 ]
```

【提示】循环体移动至最前面

实例：
```
for item in [True,"1", 1, 1.0]:
    type (item)
```

```
[ type (item) for item in [True,"1", 1, 1.0]]
```

In [33]: `print([ord(i) for i in['朝', '乐', '门']])`

Out [33]: [26397, 20048, 38376]

In [34]: 技巧 在Python列表推导式中可以用字符串的占位符，如%d，用法类似C语言的printf()、scanf()函数中的占位符。

`["input/%d.txt"% i+ "dd%d"%i for i in range(5)]`

Out [34]:
['input/0.txtdd0',
 'input/1.txtdd1',
 'input/2.txtdd2',
 'input/3.txtdd3',
 'input/4.txtdd4']

In [35]: `["input/%d.txt"%i+ "_%d" %i for i in range(5)]`

注意：%d 为占位符，在对应位置上显示的是"%i"的值。

Out [35]:
['input/0.txt_0',
 'input/1.txt_1',
 'input/2.txt_2',

```
'input/3.txt_3',
'input/4.txt_4']
```

14.7 插入与删除

In [36]:

> **提示** 可以用列表的 insert()方法向列表中新增或插入分量。
>
> ```
> lst_1 = [10,10,11,12,13,14,15]
> lst_1.insert(1, 8)
> ```
>
> **注意** 在此,"1"和"8"分别代表的是数值"8"在列表 lst_1 中的插入位置,即下标为 1 的位置。
>
> ```
> lst_1
> ```

Out [36]: [10, 8, 10, 11, 12, 13, 14, 15]

In [37]:

> **提示** 可以用列表的 pop()方法删除某个指定元素,如下标为 2 的元素。
>
> ```
> lst_1 = [10,10,11,12,13,14,15]
> lst_1.pop(2)
> lst_1
> ```

Out [37]: [10, 10, 12, 13, 14, 15]

In [38]:

> **提示** Python 支持根据下标来删除某个元素。
>
> ```
> lst_1 = [10,10,11,12,13,14,15]
> del lst_1[2]
> lst_1
> ```

Out [38]: [10, 10, 12, 13, 14, 15]

In [39]:

> **提示** Python 支持根据值来删除某个元素。

```
lst_1 = [10,10,11,12,13,14,15]
lst_1.remove(10)
```
 只删除第一个 10，第二个 10 并未删除。
```
lst_1
```

Out [39]: [10, 11, 12, 13, 14, 15]

In [40]:

 能否用以下方法删除某个值的全部元素？不一定。
```
lst_1 = [10,10,11,12,11,13,14,15]
for i in lst_1 :
    if i == 11:
```
 在此，请将"11"改为"10"，试一试？并分析结果的原因。
```
        lst_1.remove(i)
print(lst_1)
```

Out [41]: [10, 10, 12, 13, 14, 15]

In [41]:

 在 Python 中，可以用"列表推导式"进行过滤（删除）指定值。
```
lst_1 = [10,10,11,12,11,13,14,15]
[x for x in lst_1 if x != 10]
```

Out [41]: [11, 12, 11, 13, 14, 15]

In [42]:

 在 Python 中，也可以用 filter() 函数进行过滤（删除）指定值。
```
lst_1 = [10,10,11,12,11,13,14,15]
list(filter(lambda i:i != 10, lst_1))
```

Out [42]: [11, 12, 11, 13, 14, 15]

In [43]:
 技巧 | 可以用 set()函数进行重复过滤。

```
lst_1 = [10,10,11,12,11,13,14,15]
list(set(lst_1))
```

Out [43]: [10, 11, 12, 13, 14, 15]

14.8 常用操作函数

In [44]:
 提示 | 计算长度的方法：内置函数 len()。

```
len(lst_1)
```

 注意 | 此函数名为 len()，并非为 length()。

Out [44]: 8

In [45]:
 提示 | 列表的排序：内置函数 sorted()。

```
lst_1 = [10,10,11,12,11,13,14,15]
sorted(lst_1)
```

Out [45]: [10, 10, 11, 11, 12, 13, 14, 15]

In [46]:
```
lst_1
```
 注意 | 内置函数 sorted()并不改变列表本身。

Out [46]: [10, 10, 11, 12, 11, 13, 14, 15]

In [47]:
 提示 | 除了内置函数 sorted()，还可以用列表方法 sort()。

```
lst_1 = [10,10,11,12,11,13,14,15]
```

14 列表

```
lst_1.sort()
lst_1
```

 内置函数 sorted() 与列表方法 sort() 的区别：后者直接更改列表本身的内容（取值）。

Out [47]: [10, 10, 11, 11, 12, 13, 14, 15]

In [48]: 注意区分列表的方法 extend() 和 append()。

```
lst_1 = [10,10,11,12,11,13,14,15]
lst_2 = [11,12,13,14]
lst_1.append(lst_2)
```

 lst_2 作为 lst_1 的一个元素的"身份"来追加。

```
print(lst_1)
```

Out [48]: [10, 10, 11, 12, 11, 13, 14, 15, [11, 12, 13, 14]]

In [49]:
```
lst_1 = [10,10,11,12,11,13,14,15]
lst_2 = [11,12,13,14]
lst_1.extend(lst_2)
```

 lst_1 后直接追加 lst_2，即直接合并两个列表中的元素。

```
print(lst_1)
```

Out [49]: [10, 10, 11, 12, 11, 13, 14, 15, 11, 12, 13, 14]

In [50]: 列表的打印：内置函数 print()。

```
lst_1 = [1,2,3,'Python',True,4.3,None]
lst_2 = [1,2,[2,3]]
print(lst_1, lst_2)
```

Out [50]: [1, 2, 3, 'Python', True, 4.3, None] [1, 2, [2, 3]]

In [51]:
> 提示　内置函数 reversed() 与列表方法 reverse() 的区别：前者不修改列表本身，而后者直接修改列表本身。

```
lst_1 = [1,2,3,'Python',True,4.3,None]
list(reversed(lst_1))
```

> 注意　reversed(lst_1) 返回一个迭代器，需要用 list() 函数转换才能显示。

Out [51]: [None, 4.3, True, 'Python', 3, 2, 1]

In [52]:
```
reversed(lst_1)
```

> 思路　在数据分析和数据科学项目中，需要注意"操作"或"方法"是否影响（更改）被操作对象本身的值。

Out [52]: <list_reverseiterator at 0x1eba97e83c8>

In [53]:
```
lst_1
```

> 提示　reversed()：为内置函数，不改变列表本身，而是临时返回另一个反向遍历的列表。

Out [53]: [1, 2, 3, 'Python', True, 4.3, None]

In [54]:
```
lst_1 = [1,2,3,'Python',True,4.3,None]
lst_1.reverse()
lst_1
```

Out [54]: [None, 4.3, True, 'Python', 3, 2, 1]

In [55]:
> 提示　两个列表的同步计算：zip() 函数。

```
str1 = [1,2,3,4,5]
str2 = [20,21,23,24,25]
print(zip(str1,str2))
```

Out [55]: <zip object at 0x000001EBA97A9BC8>

14 列表

In [56]:
```
print(list(zip(str1,str2)))
```

 提示 zip()函数的返回值为迭代器，需进行 list()强制类型转换后才能获取其内容（取值），参见本书【25 迭代器与生成器（P165）】。

Out [56]: `[(1, 20), (2, 21), (3, 23), (4, 24), (5, 25)]`

In [57]:
 提示 与 C 和 Java 不同的是，Python 有"列表的推导式"的概念，可以用来避免复杂的 for 语句。

```
str1 = ["a","about","c","china","b","beijing"]
[x.upper() for x in str1 if len(x) > 1]
```

注意 有三个要点

```
#第一个位置（for 之前）是"将要重复计算的公式"
#第二个位置（for 和 in 之间）是"从迭代器中提取的循环变量"
#第三个位置（in 之后）是"可迭代的对象或迭代器"，本例中带有 if 语句
```

Out [57]: `['ABOUT', 'CHINA', 'BEIJING']`

In [58]:
```
[x**2 for x in range(10)]
```

 提示 列表推导式的计算顺序如下。

```
#首先计算 range(10)
#接着计算 x
#最后计算 x**2
```

Out [58]: `[0, 1, 4, 9, 16, 25, 36, 49, 64, 81]`

In [59]:
```
str1 = ["a","about","c","china","b","beijing"]
[str2.upper() for str2 in str1 if len(str2) > 1]
```

 提示 此代码中出现了 if 的三元组运算，参见本书【10 if 语句】。

Out [59]: `['ABOUT', 'CHINA', 'BEIJING']`

In [60]:
 跟踪列表的下标：内置函数 enumerate()。

```
myList = [2,3,5,6,7,3,2]
list(enumerate(myList))
```

Out [60]: [(0, 2), (1, 3), (2, 5), (3, 6), (4, 7), (5, 3), (6, 2)]

In [61]:
 在数据分析和数据科学项目中，需要注意面向软件开发的写代码与面向数据分析、数据科学的写代码之间的差异。

In [62]:
 在面向软件开发的编程时，Python 编程与 C 和 Java 编程是相似的。

```
i = 0
sum = 0
for value in myList:
    i = i + 1
    sum = value + i
sum
```

Out [62]: 9

In [63]:
 在面向数据分析、数据科学的 Python 编程时，Python 编程与 C、Java 编程差别很大，更重视的是"数据层面的问题"，而不是"计算层面的问题"。

```
sum = 0
dict((value,i) for i, value in enumerate(myList))
```

 请结合本书【19 字典】知识，解读本行代码的输出结果。

Out [63]: {2: 6, 3: 5, 5: 2, 6: 3, 7: 4}

15 元组

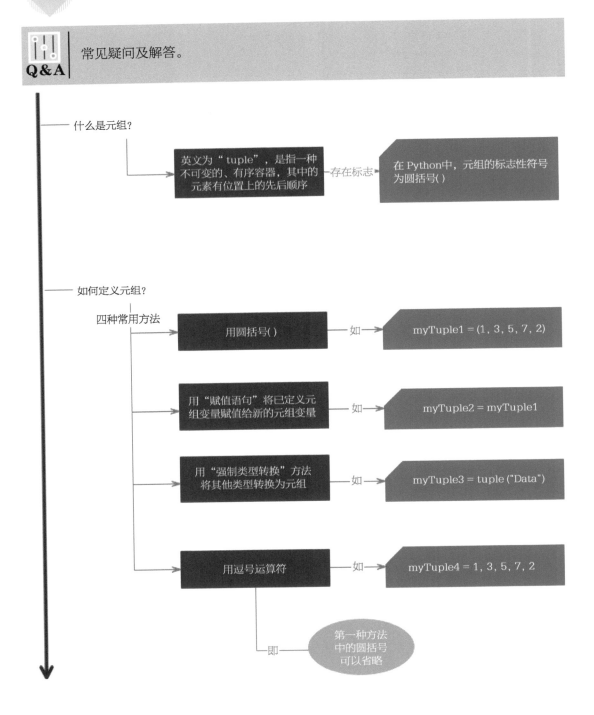

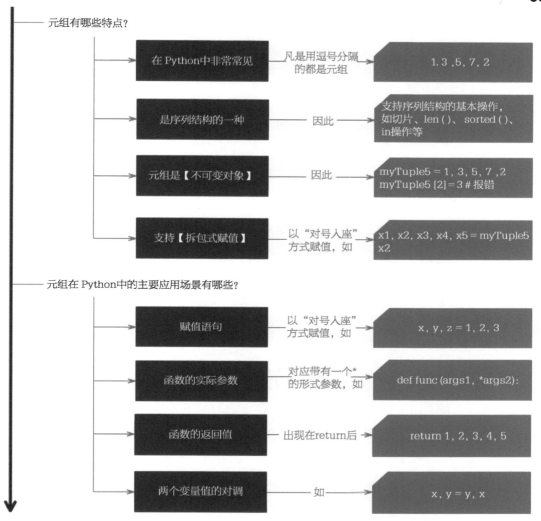

15.1 定义方法

In [1]: 第一种方法：用"圆括号+逗号"。

```
myTuple1 = (1,3,5,7,2)
print(myTuple1)
```

Out[1]: (1, 3, 5, 7, 2)

15 元组

In [2]:
 注意 | 定义元组时可以省略圆括号,但不能省略逗号。

1,3,5,7,2

 提示 | 输出结果中自动增加了圆括号。

Out[2]: (1, 3, 5, 7, 2)

In [3]:
 思路 | 第二种方法:用"赋值语句"将已定义元组的变量赋值给新元组的变量。

```
myTuple2 = myTuple1
print(myTuple2)
```

Out[3]: (1, 3, 5, 7, 2)

In [4]:
 思路 | 第三种方法:用"强制类型转换"方法,将其他类型转换为元组。

```
myTuple3 = tuple("Data")
myTuple3
```

Out[4]: ('D', 'a', 't', 'a')

In [5]:
 思路 | 第四种方法:用"逗号运算符",即第一种方法中的圆括号可以省略。

myTuple4 = 1,3,5,7,2

 注意 | 逗号运算符在Python中代表的是一个元组。

print(myTuple4)

Out[5]: (1, 3, 5, 7, 2)

15.2 主要特征

In [6]:
> 提示 | 元组在 Python 中的应用非常广泛。
>
> 1,3,5,7,2
>
> 注意 | 此处，Jupyter Notebook 在输出（Out）变量中自动增加了圆括号。

Out[6]: (1, 3, 5, 7, 2)

In [7]:
> 提示 | 元组与列表的区别：前者为"不可变对象"，后者为"可变对象"。
>
> myTuple = 1,3,5,7,2
> myTuple[2] = 100   #报错，原因："元组"为不可变对象

Out[8]:
```
TypeError                                 Traceback (most recent call last)
<ipython-input-7-2a123938adb2> in <module>()
      1 myTuple = 1,3,5,7,2
----> 2 myTuple[2] = 100 #报错，原因："元组"为不可变对象

TypeError: 'tuple' object does not support item assignment
```

In [8]:
> myList = [1,3,5,7,2]
> myList[2] = 100
>
> 提示 | 不报错，原因："列表"为可变对象。
>
> myList

Out[8]: [1, 3, 100, 7, 2]

15 元组

In [9]:
 与列表类似,元组支持切片操作,因为二者均属于"序列"类型。

```
myTuple = 1,3,5,7,2
myTuple[2:5]
```

Out[9]: (5, 7, 2)

In [10]:
 计算元组的长度:内置函数 len()。

```
myTuple = 1,3,5,7,2
len(myTuple)
```

Out[10]: 5

In [11]:
 元组的排序:内置函数 sorted()。

```
myTuple = 1,3,5,7,2
print(sorted(myTuple))
```

 sorted()函数生成了另一个新结果,新结果的数据类型是"列表",而不是"元组"。

Out[11]: [1, 2, 3, 5, 7]

In [12]:
 与列表不同,Python 的元组无 sort()方法,原因:元组为不可变对象,而 sort()方法会改变原对象本身。

```
myTuple = 1,3,5,7,2
myTuple.sort()
```

 报错,原因:元组无此方法。

Out[12]:
AttributeError   Traceback (most recent call last)

```
<ipython-input-12-6db45af06068> in <module>()
      1 myTuple = 1,3,5,7,2
----> 2 myTuple.sort()
      3      #报错,原因:元组无此方法

AttributeError: 'tuple' object has no attribute 'sort'
```

In [13]:
> 提示 | 元组的 in 操作。
>
> myTuple = 1,3,5,7,2
> 5 in myTuple

Out[13]: True

In [14]:
> 提示 | 元组的频次统计:count()方法。如 myTuple.count(11)的含义为统计 myTuple 中数值 11 的出现频次。
>
> myTuple = 1,3,5,7,2
> myTuple.count(11)

Out[14]: 0

In [15]:
> 提示 | 元组的"拆包式赋值"规则:对号入座。
>
> myTuple = 1,3,5,7,2
> x1,x2,x3,x4,x5 = myTuple
> x2

Out[15]: 3

15.3 基本用法

In [16]:
> 提示 | Python 中的特殊赋值方法——拆包式赋值。
>
> x,y,z = 1,2,3
> print(x,y,z)

Out[16]: 1 2 3

15 元组

In [17]:
 用"圆括号+逗号"表示一个元组,但圆括号可以省略。

```
myTuple = (1,5,6,3,4)
print(myTuple)
print(len(myTuple))
print(max(myTuple))
```

Out[17]:
(1, 5, 6, 3, 4)
5
6

In [18]:
 支持"拆包式赋值"。

```
myTuple = (11,12,13,12,11,11)
a1,a2,a3,a4,a5,a6=myTuple
a3
```

Out[18]: 13

In [19]:
```
myTuple = (11,12,13,12,11,11)
myTuple.count(11)
```

 频次统计,即在 myTuple 中"11"出现的频次。

Out[19]: 3

15.4 应用场景

In [20]:
 在 Python 中,元组对应的是"带有一个*的形式参数",即"元组的形参接收不定长的实参"。关于形参与实参的含义,参见本书【23.5 形参与实参(P150)】。

```python
def func(args1,*args2):
    print(args1)
    print(args2)
func("a","b","c","d","e","f")
```

Out[20]:　　a
　　　　　　('b', 'c', 'd', 'e', 'f')

In [21]:

 提示　　"*" 对应的是 "元组"，"**" 对应的是 "字典"。

```python
def func(args1,**args2):
    print(args1)
    print(args2)
func("a",x1="b",x2="c",x3="d",x4="e",x5="f")
```

注意　　字典中，实参必须显式给出 keys 值，如 x1、x2 等。

Out[21]:　　a
　　　　　　{'x1': 'b', 'x2': 'c', 'x3': 'd', 'x4': 'e', 'x5': 'f'}

In [22]:

 提示　　在 Python 中，很多函数的返回值往往为 "元组" 的原因在于，"return 1,2,3" 相当于 "return(1,2,3)"。

```python
def func():
    return 1,2,3,4,5
func()
```

Out[22]:　　(1, 2, 3, 4, 5)

In [23]:

注意　　在 Python 中，"," 代表的是 "元组"。

```
1,2
```

提示　　上一行的输出结果中自动增加了 "()"。

Out[23]:　　(1, 2)

15 元组

In [24]:

> **技巧** 在 Python 中,两个变量值的调换操作可以通过"元组"完成。
>
> ```
> x = 1
> y = 2
> x,y = y,x
> print(x,y)
> ```
>
> **提示** 更多内容参见本书【7.5 两个变量值的调换(P47/In[7])】。

Out[24]: 2 1

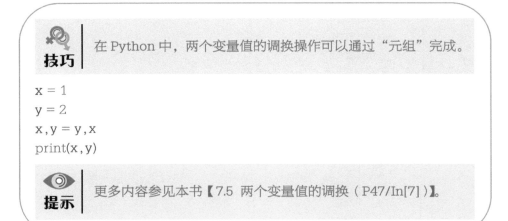

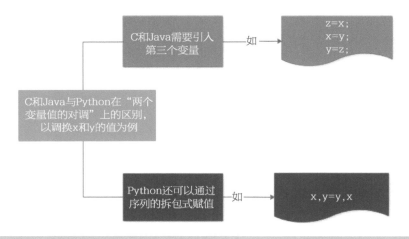

扩展 Python 中的"器"。

Python 中有很多"器"。

(1)元素之间有先后顺序、可重复、可变容器——列表,见本书【14 列表(P76)】。

(2)元素之间有先后顺序且可重复但不可变容器——元组,见本书【15 元组(P94)】。

(3)元素之间有先后顺序的容器——序列,见本书【17 序列(P111)】。

(4)元素之间无先后顺序且不可重复但可变容器——集合,见本书【18 集合(P120)】。

(5)每个元素有自己 key 的容器——字典,见本书【19 字典(P126)】。

(6)生成一个迭代器的容器(函数)——生成器,见本书【25 迭代器与生成器(P165)】。

(7)装饰器,见本书【31 面向对象编程(P197)】。

16 字符串

 常见疑问及解答。

- Python中的字符串用单引号还是用双引号？
  - 都可以。但是如果字符串本身带有单引号（双引号），字符串需要用双引号（单引号）括起来，反之亦然 —— 如 →
    - 'chaolemen'
    - "chaolemen"
    - "chao's"
    - "chao's"
  - 其实，三个单引号也可以。通常，字符串本身占多个行时，用三个单引号表示 —— 如字符串str1 →
    ```
    str1='''
    你好！
    我好
    ！
    '''
    ```

- Python中是否有转义字符？
  - 有，以"\"开始，如"\t" —— 那么，字符串本身带有"\"怎么办？
    - 第一种方法：用"\\"来代表和转义"\"
    - 第二种方法：字符串之前加一个字母r，代表原始字符串

- Python的字符串有什么特殊的地方？
  - 1 — Python中的字符串是"不可变"的对象 → Python认为"一切皆为对象"，但对象有两种：可变对象和不可变对象
  - 2 — 具有"序列"类型的共性特点，如可以用[下标]对字符串进行"切片操作" → 如，str3="chaolemen" str4=str3[1:3] → str4的值为"ha"，注意：下标的起始值为0，而不是1；切片操作是"左包含但右不包含"。因此，str4并非为"hao"

- Python中常用的字符串处理函数有哪些？
  - 字符串的合并 → 方法join() 或运算符"+"
  - 去掉字符串左右空白符 → 方法strip()
  - 计算字符串的长度 → 函数len()
  - 字符串的大小写转换 → 转换成大写：方法upper() 转换成小写：方法lower()
  - 字符串的排序 → 方法sort()
  - 判断某字符是否在字符串中 → 成员运算符in

16 字符串

16.1 定义方法

 提示 与 C 和 Java 不同的是，Python 中统一了"字符"与"字符串"的概念，不再区别对待二者，"字符"是"字符串"的一种特例。

In [1]:
 提示 Python 字符串可以用单引号括起来，也可以用双引号括起来。

```
print('abc')
print("abc")
```

Out[1]　abc
　　　　abc

In [2]:
 提示 当字符串本身含有单引号时，字符串只能用双引号括起来，反之亦然。

```
print("abc'de'f")
```

 注意 此时，print()的输出都是单引号。

Out[2]　abc'de'f

In [3]:
```
print('abc"de"f')
```

 注意 此时，print()的输出都是双引号。

Out[3]　abc"de"f

In [4]:
 提示 Python 中也可以使用三个引号，其功能是表示"带有换行的字符串"。

```
str1 = '''
    你好!
    我好
 !
 '''
```

> 提示 | 更多内容参见本书【28 帮助文档（P180）】。

Out[4]:   '\n 你好！\n 我好\n !\n'

16.2 主要特征

In [5]:
> 提示 | 特征之一：Python 中的字符串是"不可变对象"。
>
> str1[0:4] = "2222"
> #报错 TypeError: 'str' object does not support item assignment

Out[5]:
```
TypeError                                 Traceback (most recent call last)
<ipython-input-5-cb4f88cdc030> in <module>()
      1
----> 2 str1[1:4] = "2222"
      3 #报错 TypeError: 'str' object does not support item assignment

TypeError: 'str' object does not support item assignment
```

In [6]:
> str1 = "abc"
> str1 = "defghijk"
>
> 提示 | 上行代码的执行不会报错，原因：Python 是动态类型语言，参见本书【5.2 Python 是动态类型语言（P29/In[2]）】。
>
> 注意 | "不可变对象"的含义为该对象的内容（取值）不会发生局部改变，与"动态类型语言"是不同概念。
>
> str1[1:4]

Out[6]:   'efg'

In [7]:
> 提示 | 特征之二：Python 中的字符串属于"序列"。
>
> #凡是支持序列结构的运算符和函数都可以用于字符串
> #如 Python 字符串支持"切片操作"

16 字符串

In [8]: 字符串支持"切片操作"。注意:切片操作的规则为"左包含但右不包含",即在此代码中,切片后的结果中包含下标为 0 的元素,但不包含下标为 2 的元素。

```
'clm'[0:2]
```

Out [8]: 'cl'

In [9]:
```
str3 = "chaolemen"
str4 = str3[1:3]
str4
```

Out [9]: 'ha'

In [10]: `"chaolemen"[:6]`

Out [10]: 'chaole'

16.3 字符串的操作

In [11]: 字符串合并。

```
'-'.join(['c', 'l'])
#在上一行代码中,符号"_"为字符串连接符
```

Out [11]: 'c-l'

In [12]: `'c' + 'lm'`

Out [12]: 'clm'

In [13]: 去掉字符串的左右空白符,如空格、换行符等。

```
" chaolemen ".strip()
```

Out [13]: 'chaolemen'

In [14]: 判断一个字符(串)是否在另一个字符串中。

`'c' in'clm'`

Out [14]: True

In [15]:
 计算字符串的长度。

```
len('clm')
```

Out [15]: 3

In [16]:
 计算字符的 Unicode 编码：内置函数 ord()。目前，Python 所采用的字符集编码方式为 UTF-8（8-bit Unicode Transformation Format）。

```
print(ord('A'))
print(chr(97))
```

 内置函数 chr()的功能与内置函数 ord()的相反，显示 Unicode 编码对应的字符。注意：函数 chr()的名称并不是 char。

Out [16]: 65
a

In [17]:
```
print(ord('朝'))
print(chr(26397))
```

 查看 Python 字符集的方法。

```
import sys
sys.getdefaultencoding()
```

Out [17]: 26397
朝

In [18]:
 转义字符。

```
s = 'a\tbbc'
s
```

Out [18]: 'a\tbbc'

16 字符串

In [19]:

 当字符串中含有"转义字符"时,两种输出方法"s"和"print(s)"的区别在于,前者不做转义。

print(s)

Out [19]: a bbc

In [20]:

 用 str()函数将其转换为字符串。

str(1234567)

Out [20]: '1234567'

In [21]:

 大小写转换。
大写转换为小写用方法 lower(),反之用方法 upper()。

"abc".upper()

Out [21]: 'ABC'

In [22]:

 特殊字符及路径问题。

s1 = "E:\SparkR\My\T"
s1

 在 Jupyter Notebook 中,直接显示字符串 s1 与通过 Python 内置函数 print()即 print(x)语句的输出结果不同。

Out [22]: 'E:\\SparkR\\My\\T'

In [23]:

s1 = r"http://www.chaolemen.org"
s1

 r 代表的是原始字符串,原始字符串中不会转义,即不解释转义字符,如并不将"\n"解释成"回车符"。因此,在数据分析中,原始字符串常用于文件路径的表示。

 有时,在 Python 中会遇到以"u/U"开头的字符串,那是 Python 2 的语法,Python 3 中不需要。

> **注意**：Python 2 中有两种字符串类型：Unicode 字符串和非 Unicode 字符串。Python 3 中只有一种字符串类型：Unicode 字符串。

Out [23]: 'http://www.chaolemen.org'

In [24]:
> **提示**：字符串的 join 操作。

```python
sep_str = "-"
seq = ("a", "b", "c")    # 字符串序列
sep_str.join(seq)
```

> **注意**：join() 方法的参数为"序列"，"."之前的变量（此处为 seq_str）为分隔符。

Out [24]: 'a-b-c'

In [25]:
> **提示**：字符串的排序。

```python
str1 = ["abc","aaba","adefg","bb","c"]
str1.sort()
str1
```

Out [25]: ['aaba', 'abc', 'adefg', 'bb', 'c']

In [26]:
> **提示**：按长度排序。

```python
str1.sort(key = lambda x:len(list(x)))
str1
```

Out [26]: ['c', 'bb', 'abc', 'aaba', 'adefg']

In [27]:
> **提示**：按包含的不同字符的个数排序。

16 字符串

```
str1.sort(key = lambda x:len(set(x)))
str1
```

Out [27]: ['c', 'bb', 'aaba', 'abc', 'adefg']

In [28]:
 提示 | Python 字符串支持"强制类型转换"。

```
print("str = ", str1)
print("list(str1) = ", list(str1))    #转换成列表
```

Out [28]: str = ['c', 'bb', 'aaba', 'abc', 'adefg']
list(str1) = ['c', 'bb', 'aaba', 'abc', 'adefg']

In [29]:
 提示 | 强制类型转换成集合——set()函数。

```
print("set(str1) = ", set(str1))
```

Out [29]: set(str1) = {'adefg', 'c', 'abc', 'aaba', 'bb'}

In [30]:
 思路 | Python 正则表达式（Regular Expression）的功能可以用模块 re 实现。

```
import re
p1 = re.compile('[a-dA-D]')
r1 = p1.findall('chaolemen@ruc.edu.cn')
r1
```

 提示 | 函数 compile()和方法 findall()的功能分别为"自定义一个正则表达式"和"用自定义正则表达式匹配目标字符串中的所有子字符串"。

 思路 | Python 正则表达式的语法可以参见官方文档：https://docs.python.org/2/library/re.html。

Out [30]: ['c', 'a', 'c', 'd', 'c']

17 序列

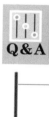

常见疑问及解答。

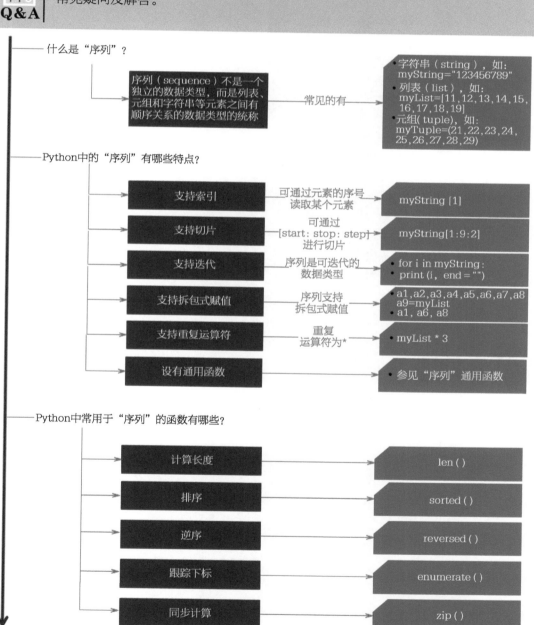

17 序列

In [1]:
> **提示** 如何正确理解 Python 中的"序列"?

```
#"序列"的含义
    #有序的元素集合
#常见类型
    #列表、元组和字符串等三种。但是注意：集合（Set）不属于
    #序列类型
#主要特点
    #支持索引
    #支持切片
    #可以迭代
    #支持拆包
    #支持*运算
    #设有共同的内置函数
#常用函数
    #enumerate
    #reversed
    #sorted
```

17.1 支持索引

In [2]:
> **提示** 可以通过索引/下标读取某个元素。

In [3]:
```
myString = "123456789"
myString[1]
```

Out [3]: '2'

In [4]:
```
myList = [11,12,13,14,15,16,17,18,19]
myList[1]
```

Out [4]: 12

In [5]:
```
myTuple = (21,22,23,24,25,26,27,28,29)
myTuple[1]
```
Out [5]: 22

17.2 支持切片

In [6]:
 提示　可以通过[start:stop:step]进行切片,参见本书【14 列表 (P80/In[8])】。

In [7]:
```
myString = "123456789"
myString[1:9:2]
```
Out [7]: '2468'

In [8]:
```
myList = [11,12,13,14,15,16,17,18,19]
myList[1:9:2]
```
Out [8]: [12, 14, 16, 18]

In [9]:
```
myTuple = (21,22,23,24,25,26,27,28,29)
myTuple[1:9:2]
```
Out [9]: (22, 24, 26, 28)

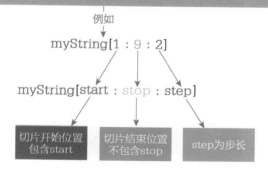

17.3 支持迭代

In [10]:

"序列"是可迭代的数据类型,可以放在 for 语句的 in 之后。

```
myString = "123456789"
for i in myString:
    print(i, end = " ")
```

Out [10]: 1 2 3 4 5 6 7 8 9

In [11]:
```
myList = [11, 12, 13, 14, 15, 16, 17, 18, 19]
for i in myList:
    print(i, end = " ")
```

Out [11]: 11 12 13 14 15 16 17 18 19

In [12]:
```
myTuple = (21, 22, 23, 24, 25, 26, 27, 28, 29)
for i in myTuple:
    print(i, end = " ")
```

Out [12]: 21 22 23 24 25 26 27 28 29

17.4 支持拆包

In [13]:

序列支持"拆包式赋值",有时称为"并行赋值"。
赋值规则:"对号入座"。

```
myString = "123456789"
a1, a2, a3, a4, a5, a6, a7, a8, a9 = myString
a1, a2, a3, a4, a5, a6, a7, a8, a9
```

Out [13]: ('1', '2', '3', '4', '5', '6', '7', '8', '9')

In [14]:
```
myList = [11,12,13,14,15,16,17,18,19]
a1,a2,a3,a4,a5,a6,a7,a8,a9 = myList
a1,a2,a3,a4,a5,a6,a7,a8,a9
```

Out [14]: (11, 12, 13, 14, 15, 16, 17, 18, 19)

In [15]:
```
myTuple = (21,22,23,24,25,26,27,28,29)
a1,a2,a3,a4,a5,a6,a7,a8,a9 = myTuple
a1,a2,a3,a4,a5,a6,a7,a8,a9
```

Out [15]: (21, 22, 23, 24, 25, 26, 27, 28, 29)

17.5 支持*运算

In [16]:

 提示 | 序列的重复运算符为*。

```
myString = "123456789"
myString * 3
```

 注意 | 序列的*运算为"重复运算",并不是"乘法"。

Out [16]: '123456789123456789123456789'

In [17]:
```
myList = [11,12,13,14,15,16,17,18,19]
myList * 3
```

Out [17]:
[11,
 12,
 13,
 14,
 15,
 16,
 17,
 18,
 19,

17 序列

```
     11,
     12,
     13,
     14,
     15,
     16,
     17,
     18,
     19,
     11,
     12,
     13,
     14,
     15,
     16,
     17,
     18,
     19]
```

In [18]:
```
myTuple = (21,22,23,24,25,26,27,28,29)
myTuple * 3
```

Out[18]:
```
(21,
 22,
 23,
 24,
 25,
 26,
 27,
 28,
 29,
 21,
 22,
 23,
 24,
 25,
 26,
 27,
```

       28,
       29,
       21,
       22,
       23,
       24,
       25,
       26,
       27,
       28,
       29)

17.6 通用函数

思路 | 在 Python 中，凡是"序列"类型的对象，不管属于什么数据类型（如列表、元组、字符串等），均支持共性函数。

In [19]:
```
myString = "123456789"
myList = [11,12,13,14,15,16,17,18,19]
myTuple = (21,22,23,24,25,26,27,28,29)
```

提示 | 计算长度：len()。

len(myString),len(myList),len(myTuple)

Out [19]: (9, 9, 9)

In [20]:

提示 | 排序：sorted()。

sorted(myString),sorted(myList),sorted(myTuple)

Out[20]: (['1', '2', '3', '4', '5', '6', '7', '8', '9'],
          [11, 12, 13, 14, 15, 16, 17, 18, 19],
          [21, 22, 23, 24, 25, 26, 27, 28, 29])

17 序列

In [21]:
> **提示** 逆序:reversed()。
>
> reversed(myString), reversed(myList), reversed(myTuple)

Out[21]: (<reversed at 0x209a35bec50>,
 <list_reverseiterator at 0x209a35be9e8>,
 <reversed at 0x209a35be518>)

In [22]:
> **注意** reversed 返回的是迭代器,支持惰性计算,可以用内置函数 list() 将其转换为列表,参见本书【25 迭代器与生成器(P165)】。
>
> list(reversed(myString))

Out[22]: ['9', '8', '7', '6', '5', '4', '3', '2', '1']

In [23]:
> **提示** 跟踪和枚举下标:enumerate()。
>
> enumerate(myString), enumerate(myList), enumerate(myTuple)

Out[23]: (<enumerate at 0x209a35ceb88>,
 <enumerate at 0x209a35ce168>,
 <enumerate at 0x209a35cebd0>)

In [24]:
> **注意** enumerate 返回的是迭代器,可以用 list()转换为列表。
>
> list(enumerate(myString))

Out[24]: [(0, '1'),
 (1, '2'),
 (2, '3'),
 (3, '4'),
 (4, '5'),
 (5, '6'),
 (6, '7'),
 (7, '8'),
 (8, '9')]

In [25]:

提示 | 两个对象的同步计算：zip()。

zip(myList, myTuple)

zip()原理示意图

Out[25]: <zip at 0x209a252e048>

In [26]:

注意 | 内置函数 zip()返回的是迭代器，可以用另一个内置函数 list()将其转换为列表，参见本书【25 迭代器与生成器（P165）】

list(zip(myList,myTuple))

Out [26]:
```
[(11, 21),
 (12, 22),
 (13, 23),
 (14, 24),
 (15, 25),
 (16, 26),
 (17, 27),
 (18, 28),
 (19, 29)]
```

提示 | 与列表、元组、集合和字典等不同，序列并不是 Python 的独立数据类型，而是包含列表、元组和字符串在内的多个数据类型的统称。

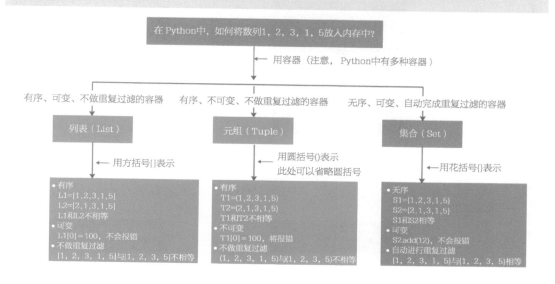

18 集合

 常见疑问及解答。

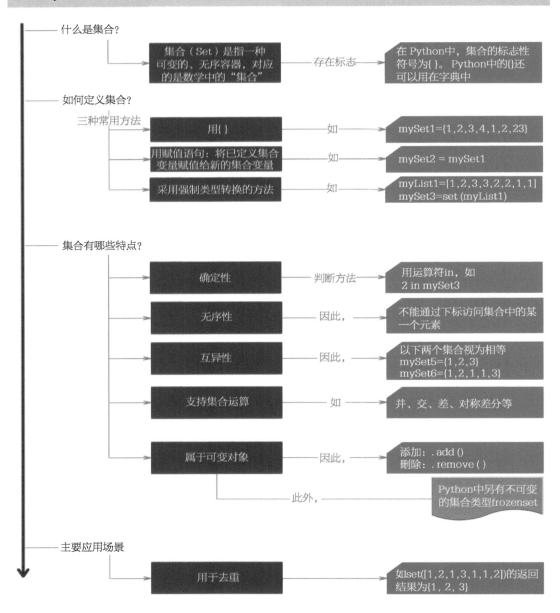

18.1 定义方法

In [1]:
>
> **提示** 第一种方法：直接定义集合（set）变量的方法，用{}。

```
mySet1 = {1,2,3,4,1,2,23}
mySet1
#从该定义方法可以看出，集合（Set）类型的本质就是"只有值（value）而没有键（key）的无序数据结构"
```

Out [1]: {1, 2, 3, 4, 23}

In [2]:
>
> **提示** 第二种用法：用赋值语句，即将"已定义集合变量"赋值给"新的集合变量"。

```
mySet2 = mySet1
mySet2
```

Out [2]: {1, 2, 3, 4, 23}

In [3]:
>
> **提示** 第三种方法：用 set()函数进行"强制类型转换"，即将"其他类型的对象"转换为"集合对象"。

```
myList1 = [1,2,3,3,2,2,1,1]
mySet3 = set(myList1)
mySet3
```

Out [3]: {1, 2, 3}

In [4]:
```
#再如
mySet4 = set("chaolemen")
mySet4
```

Out [4]: {'a', 'c', 'e', 'h', 'l', 'm', 'n', 'o'}

18 集合

18.2 主要特征

In [5]:
> **提示** 确定性：给定一个集合，任给一个元素，该元素或者属于或者不属于该集合，二者必居其一，不允许有模棱两可的情况出现。
>
> 2 in mySet3

Out [5]: True

In [6]:
> **提示** 无序性：集合中的元素是无序的。因此，在Python中不能用下标访问集合中的元素。
>
> mySet4[2]
>
> **注意** 报错 TypeError: 'set' object does not support indexing。
> 原因分析：集合具有无序性，没有下标。

Out [6]:
```
TypeError                                 Traceback (most recent call last)
<ipython-input-6-cbe34bb7c0c2> in <module>()
      1 #无序性
      2 #集合中的元素是无序的，所以不能用下标访问集合中的元素
----> 3 mySet4[2] #报错 TypeError: 'set' object does not support indexing

TypeError: 'set' object does not support indexing
```

In [7]:
> **提示** 互异性：集合中的元素是互异的。因此，在Python中，以下两个集合被视为相等。
>
> mySet5 = {1,2,3}
> mySet6 = {1,2,1,1,3}
> mySet5 == mySet6

Out [7]: True

18.3 基本运算

In [8]:
```
mySet7 = {1,3,5,10}
mySet8 = {2,4,6,10}
```

In [9]:
```
#包含
3 in mySet7
```

Out [9]: True

In [10]:
```
#不包含
3 not in mySet7
```

Out [10]: False

In [11]:
```
#等于
mySet7 == mySet8
```

Out [11]: False

In [12]:
```
#不等于
mySet7 != mySet8
```

Out [12]: True

In [13]:
```
#子集
{1,5} < mySet7
```

Out [13]: True

In [14]:
```
#合集
mySet7 | mySet8
```

Out [14]: {1, 2, 3, 4, 5, 6, 10}

In [15]:
```
#交集
mySet7 & mySet8
```

Out [15]: {10}

18 集合

In [16]:
```
#差集
mySet7 - mySet8
```

Out [16]: {1, 3, 5}

In [17]:
```
#对称差分
mySet7 ^ mySet8
```

Out [17]: {1, 2, 3, 4, 5, 6}

In [18]:
```
#判断是否为子集
print({1,3}.issubset(mySet7))
```

Out [18]: True

In [19]:
```
#判断是否为父集
print({1,3,2,4}.issuperset(mySet7))
```

Out [19]: False

In [20]:

 提示 | 在 Python 中，集合分为 set 和 frozenset 两种。set 是可变对象。

```
mySet9 = {1,2,3,4}
mySet9.add(4)
mySet9.remove(1)
mySet9
```

Out [20]: {2, 3, 4}

In [21]:

 提示 | frozenset 是不可变对象。

```
mySet10 = frozenset({1,2,3,4})
mySet10
```

 注意 | 在数据分析和数据科学项目中，为了保护数据，防止数据在分析过程中被修改，通常采用不可变对象技术。

Out [21]: frozenset({1, 2, 3, 4})

In [22]: mySet10.add(5)

> 注意 | 报错。原因：frozenset 是不可变对象。

Out [22]:
```
AttributeError                            Traceback (most recent call last)
<ipython-input-22-69d176bcbb99> in <module>()
----> 1 mySet10.add(5) #报错：AttributeError: 'frozenset' object has no attribute 'add'

AttributeError: 'frozenset' object has no attribute 'add'
```

18.4 应用场景

In [23]:
 提示 | 由于集合具有"互异性"特点，在数据分析或数据科学项目中，通常用于"去重处理"。

```
myList = ["d","a","t","a"]
mySet11 = set(myList)
mySet11
```

Out [23]: {'a', 'd', 't'}

 扩展 | Python 中支持"单行写法"。

Python 中支持"单行写法"，比较常见的包括如下内容。
（1）if 语句的单行写法——三元运算符，参见本书【10.3 if 与三元运算（P63/In[4]）】。
（2）for 语句的单行写法——列表推导式，参见本书【14.6 列表推导式（P84/In[27]）】。
（3）函数的单行写法——lambda 函数，参见本书【24 lambda 函数（P158）】。

19 字典

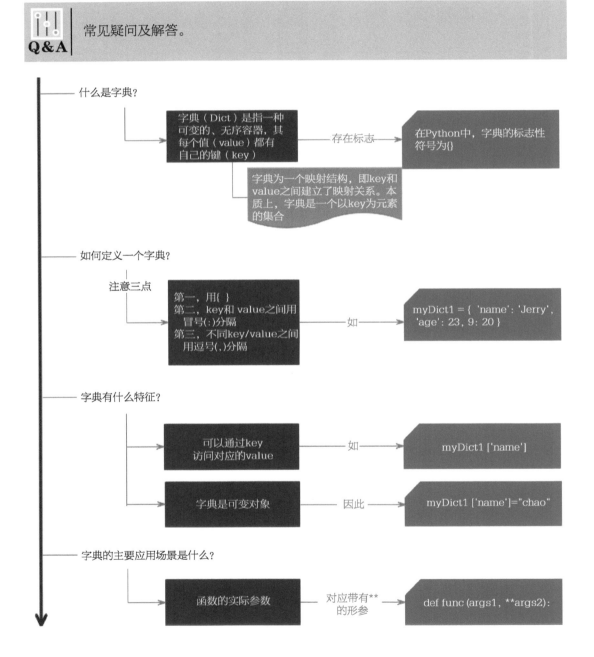

19.1 定义方法

In [1]:

提示 字典（Dict）是一种映射结构，是指一种无序容器，其中的每个元素（value）都有自己的 key。

提示 Python 中的字典相当于"R 语言中的 list"。

In [2]:

思路 定义一个字典的方法：用{ }。

```
myDict1 = {'name': 'Jerry', 'age': 23,9:20}
myDict1
```

注意 定义一个字典时应注意的事项。

```
# 第一，用{ }。
# 第二，key 与 value 之间用 ":" 分隔
# 第三，不同 key/value 之间用 "," 分隔
```

Out[2]: {9: 20, 'age': 23, 'name': 'Jerry'}

In [3]:

提示 如果两个 values 的 key 相同，会发生什么？请看 myDict2 的输出结果。

```
myDict2 = {2:2,2:3,4:5}
myDict2
```

Out[3]: {2: 3, 4: 5}

注意 "字典"是无序结构，访问其中的 value（元素值）时不能通过下标，而应采用 key（键）。key 是区分 value 的唯一依据。

19.2 字典的主要特征

In [4]:
> 提示　特征一：可以通过 key（键）访问对应的 value（值）。

```
myDict1['name']
```

> 注意　当 key 为字符串时，需要用双引号括起来，否则报错：
> NameError: name 'a' is not defined

Out[4]: 'Jerry'

In [5]:
```
myDict1[9]
```

> 提示　此处，key 为整数（9），而不是字符串。

> 思路　可见，字典类对象的下标中可以写 key。

Out[5]: 20

In [6]:
> 提示　特征二：字典是可变对象。

```
myDict1 = {'name': 'Jerry', 'age': 23,9:20}
myDict1['name'] = "chao"
myDict1
```

Out[6]: {9: 20, 'age': 23, 'name': 'chao'}

In [7]:
> 注意　字典的 key 不能为"不可哈希"（unhashable type）对象，Python 的可变数据对象（如列表、集合、字典等）为不可哈希对象。

```
dct3 = {[2,3]:[4,4], 5:5}
```

> 提示　出错，提示为 TypeError: unhashable type: 'list'。
> 原因分析：key = [2,3]为不可哈希对象。Python 语言中的可变对象是不可哈希对象。

Out [7]:
```
TypeError                                 Traceback (most recent call last)
<ipython-input-8-f53f76c22a27> in <module>()
----> 1 dct3={[2,3]:[4,4], 5:5}

TypeError: unhashable type: 'list'
```

19.3 字典的应用场景

In [8]:

 提示 | 字典在数据分析/数据科学中的主要应用场景是存放"临时数据",如函数参数"**args"。

 注意 | 在函数形参中,带有"*"和"**"的参数分别代表的是以"元组(不带 key 的值)"和"字典(带 key 的值)"的形参接收不定长的实参。

```
def func(args1,**args2):
    print(args1)
    print(args2)
func("a",x1="b",x2="c",x3="d",x4="e",x5="f")
```

 提示 | 当实参为"字典"时,实参中必须以显式方式指定对应的 key(键)。

Out [8]:   a
{'x1': 'b', 'x2': 'c', 'x3': 'd', 'x4': 'e', 'x5': 'f'}

 扩展 | "字典"数据类型。

本章介绍了"字典"数据类型,是一种 key-value 结构。NoSQL 数据库是大数据分析和数据科学项目中常用的技术,其基本数据模型也是 key-value 结构。根据 value 的不同,NoSQL 数据库可以分为 key-value 型、key-document 型和 key-column 型等数据模型,参见本书【47 基于 Spark 和 MongoDB 的大数据分析(P449)】。

20 函数

 常见疑问及解答。

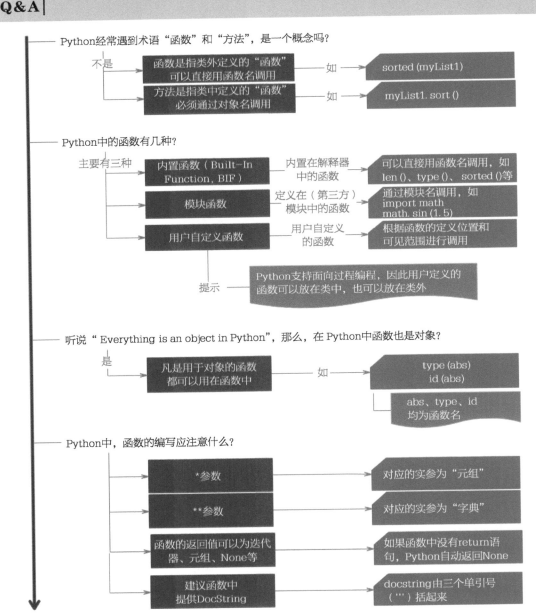

In [1]:
 思路 | Python 的函数可以分为内置函数、模块函数和用户自定义函数三类。

```
#第一，用户自定义函数可以写成单行函数，称为"lambda 函数"
#第二，用户自定义函数可以放在"类（class）"中，也可以放在类外。原
#因在于，Python 不仅支持面向对象编程，还支持面向过程编程
```

20.1 内置函数

In [2]:
 提示 | 内置函数（Built-In Function，BIF）是指已内置在 Python 解释器中的函数，其调用方法为"直接用函数名"。

```
i=20
type(i)
```

Out[2]: int

In [3]:
 思路 | 更多知识请参见本书【21 内置函数（P133）】。

20.2 模块函数

In [4]:
 提示 | 模块函数是指定义在 Python 模块中的函数，其调用方法为"先 import 所属模块，后通过'模块名'或'模块别名'调用"。

```
import math as mt
mt.sin(1.5)
```

Out[4]: 0.9974949866040544

In [5]:
 思路 | 更多知识请参见本书【22 模块函数（P141）】。

20.3 用户自定义函数

In [6]:

提示　"用户自定义函数"是指我们（用户）自己定义的函数，其调用方法为"直接用函数名"。

思路　用关键字 def 定义"用户自定义函数"。

```
def myFunc():
    j = 0
    print('hello world')
myFunc()
```

Out [6]:　hello world

In [7]:

思路　有关更多知识,参见本书【23 自定义函数（P145）】。

提示　在 Python 中，用户自定义函数可以写成"单行函数"，称为"lambda 函数"，参见本书【24 lambda 函数（P158）】。

注意　Python 既支持面向对象编程，又支持面向过程编程。因此，用户自定义函数可以放在类中，也可以放在类外。前者被称为"方法"，后者被称为"函数"。初学者需要区分"函数"和"方法"的概念，参见本书【31 面向对象编程（P199/In[2]）】。

开发自己的第三方扩展包。

　　Python 既支持函数式编程（Functional Programming），又支持面向对象编程（Object Oriented Programming），方便了不同数据分析和数据科学项目的完成。通常，当调用第三方包来完成自己的数据分析项目时，用函数式编程方法更为方便、实用。但是，当开发一个新的第三方包时，建议使用面向对象方法，以便程序代码的可复用和可维护。

　　关于如何开发自己的第三方扩展包,建议查阅 Python 官网文档 How To Package Your Python Code（https://python-packaging.readthedocs.io/en/latest/）。

21 内置函数

Q&A 常见疑问及解答。

—— 什么是"内置函数"？
→ 指内置在Python解释器中的函数，可以直接通过函数名（不需要提供所属模块名或类名的前提下）调用的函数 —— 如 → len(myList1)

—— 如何查看内置函数的说明或帮助？
→ help()函数或"?"操作符 —— 如 → help(len)或len?
→ 通常，Python帮助提示中给出的信息有限，建议读者养成阅读官网文档的习惯

—— 常用内置函数有哪些？
→ 有很多 —— 可分为 → 数学函数 / 类型函数 / 其他功能函数等

- help()：查看帮助
- type()：查看类型
- id()：查看id
- len()：计算长度
- isinstance()：判断是否属于特定数据类型

21 内置函数

21.1 内置函数的主要特点

In [1]:

 内置函数（BIF，Built-In Function）的主要特点是"直接通过函数名调用"，如 type() 函数。

i = 20
type(20)

 在 Python 中，"函数"与"方法"是两个不同的概念。

 为了提升性能，Python 中的很多内置函数并不是用 Python 编写的，而是用 C 语言等编写的。也就是说，Python 内置函数的源代码不一定是 Python 代码。

Out [1]:　int

 查看内置函数的方法——内置函数 dir()。

dir(__builtins__)

21.2 数学函数

In [2]:

abs(-1)

 求绝对值。

Out [2]:　1

In [3]:

min([1,2,3])

 求最小值。

Out [3]:　1

In [4]: max([1,2,3])

> 提示 ｜ 求最大值。

Out [4]: 3

In [5]: pow(2,10)

> 提示 ｜ 求 2 的 10 次方。

Out [5]: 1024

In [6]: round(2.991,2)

> 提示 ｜ 四舍五入，第二个参数的含义为小数点后保留的位数。

Out [6]: 2.99

21.3 类型函数

In [7]:
> 提示 ｜ 强制类型转换为 int（整型）：
> int()

int(1.134)

> 思路 ｜ 一般情况下，Python 中的强制类型转换时所需的函数名与目标数据类型的名称一致。

Out [7]: 1

In [8]:
> 提示 ｜ 强制类型转换为 bool（布尔类型）：
> bool()

bool(1)

Out [8]: True

21 内置函数

In [9]:
> 提示 强制类型转换为 float（浮点型）：
> float()

float(1)

Out [9]: 1.0

In [10]:
> 提示 强制类型转换为 str（字符型）：
> str()

str(123)

Out [10]: '123'

In [11]:
> 提示 强制类型转换为 list（列表）：
> list()

list("chao")

Out [11]: ['c', 'h', 'a', 'o']

In [12]:
> 提示 强制类型转换为 set（集合）：
> set()

set("chao")

Out [12]: {'a', 'c', 'h', 'o'}

In [13]:
> 提示 强制类型转换为 tuple（元组）：
> tuple()

tuple("chao")

Out [13]: ('c', 'h', 'a', 'o')

21.4 其他功能函数

In [14]:
> 提示 查看数据类型：
> type()

i = 0
type(i)

Out [14]: int

In [15]:
 判断数据类型：
isinstance()

isinstance(i, int)

Out [15]: True

In [16]:
 查看变量的搜索路径：
调用内置函数 dir()或用魔术命令%whos 和%who

dir()

Out [16]: ['In',
'Out',
'_',
'_1',
'_10',
'_11',
'_12',
'_13',
'_14',
'_15',
'_2',
'_3',
'_4',
'_5',
'_6',
'_7',
'_8',
'_9',
'__',
'___',
'__builtin__',
'__builtins__',
'__doc__',
'__loader__',
'__name__',
'__package__',
'__spec__',

21 内置函数

```
 '_dh',
 '_i',
 '_i1',
 '_i10',
 '_i11',
 '_i12',
 '_i13',
 '_i14',
 '_i15',
 '_i16',
 '_i2',
 '_i3',
 '_i4',
 '_i5',
 '_i6',
 '_i7',
 '_i8',
 '_i9',
 '_ih',
 '_ii',
 '_iii',
 '_oh',
 'exit',
 'get_ipython',
 'i',
 'quit']
```

In [17]:

 查看帮助：
help()

```
help(dir)
```

Out [17]: Help on built-in function dir in module builtins:

dir(...)
    dir([object]) -> list of strings
    If called without an argument, return the names in the current scope .
    Else, return an alphabetized list of names comprising (some of) theattributes
    of the given object, and of attributes reachable from it.

If the object supplies a method named __dir__, it will be used; otherwise
the default dir() logic is used and returns:
for a module object: the module's attributes.
for a class object: its attributes, and recursively the attribute s of its bases.
for any other object: its attributes, its class's attributes, and recursively the attributes of its class's base classes.

In [18]:

提示 | 计算长度：len()

```
myList = [1,2,3,4,5]
len(myList)
```

Out [18]: 5

In [19]:

提示 | 快速生成序列：range()

```
range(1,10,2)
```

提示 | range(1,10,2)用于生成一个迭代器，其起始位置（start）为 1（注：含 1），结束位置（stop）为 10（注：不含 10），步长为 2。参见本书【14 列表（P76）】。

Out [19]: range(1, 10, 2)

In [20]:

```
list(range(1,10,2))
```

提示 | range()的返回值为一个迭代器，迭代器是"惰性计算"的，所以通过强制类型转换函数 list()将迭代器的值进行计算并显示。参见本书【25 迭代器与生成器（P165）】。

Out [20]: [1, 3, 5, 7, 9]

In [21]:

提示 | 判断函数可否被调用：callable()

```
callable(dir)
```

Out [21]: True

21 内置函数

In [22]:
 十进制转换为二进制：bin()

```
bin(8)
```

Out [22]: '0b1000'

In [23]:
 十进制转换为十六进制：hex()

```
hex(8)
```

Out [23]: '0x8'

 Python 及其第三方包的特点。

　　Python 及其第三方包中的一些编程理念较好地支持数据分析和数据科学项目的新需求（区别于软件开发的特殊需求），如下所述。
　　（1）交互式编程与解释性语言技术，参见本书【3 如何看懂和运行本书代码？（P8）】。
　　（2）强类型语言技术，参见本书【5.3 Python 是强类型语言（P30/In[3]）】。
　　（3）动态类型语言技术，参见本书【5.2 Python 是动态类型语言（P29/In[2]）】。
　　（4）显式索引技术，参见本书【38 DataFrame（P278/In[5]）】。
　　（5）鸭子类型（Duck Typing）编程，参见本书【28.5 dir()函数（P185/In[13]）】。
　　（6）ufunc 编程与向量化计算，参见本书【36.7 ndarray 的计算（P258/In[69]）】。
　　（7）广播机制，参见本书【36.11 ndarray 的广播规则（P261/In[81]）】。
　　（8）惰性计算，参见本书【25 迭代器与生成器（P168/In[5]）】。
　　（9）数据保护及就地修改机制，参见本书【38 DataFrame（P287/In[34]）】。
　　（10）切片与列表推导式方法，参见本书【14.6 列表推导式（P84/In[27]）】。

22 模块函数

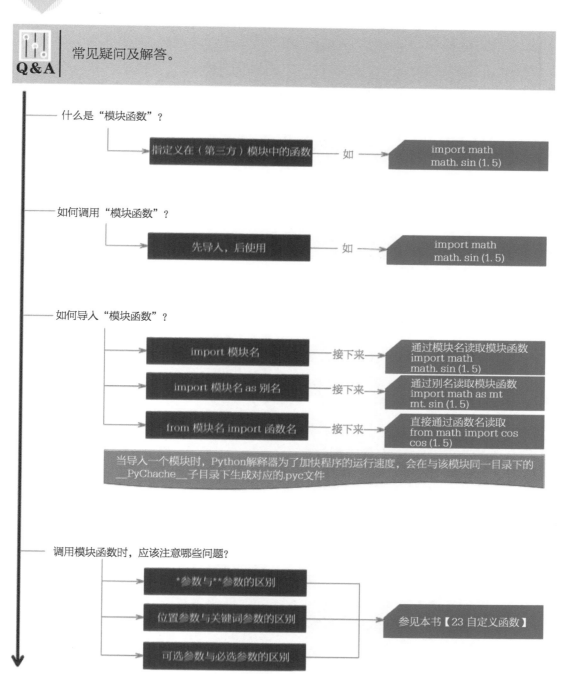

22 模块函数

In [1]:

思路 与"内置函数"不同的是,"模块函数"的定义在第三方提供的包或模块中,其调用必须在"导入模块"的前提下进行,并通常通过模块名调用。

注意 Python 中模块的导入方法有多种,不同的导入方法对应的函数调用方法不同。

🔍 **提醒** 在第三方包/模块的导入前,需要用 PIP 工具(或 Conda 工具)将其从 PyPi(或 Conda)服务器下载,参见本书【27 包(P175/In[2])】和【26 模块(P170/In[1])】。但是,为了方便用户的编程,在 Jupyter Notebook 中已经预安装了数据分析和数据科学中的常用包,所以本书中包的导入之前一般不需要下载和安装,除了【46 Spark 编程(P425)】中的 PySpark 等特殊包。

22.1 import 模块名

In [2]:

提示 第一种导入方法:
import 模块名

注意 函数的调用方法:
模块名.函数名()

```
import math
math.sin(1.5)
```

Out [2]: 0.9974949866040544

In [3]: cos(1.5)

提示 报错,NameError: name 'cos' is not defined。报错原因分析:在调用 cos() 函数时,没有给出其模块名"math"。

Out [3]:
```
---------------------------------
NameError                                 Traceback (most recent call last)
```

```
<ipython-input-2-25c1f7923f0b> in <module>()
----> 1 cos(1.5) #报错,NameError: name 'cos' is not defined

NameError: name 'cos' is not defined
```

In [4]:
> 纠正方法:
> 模块名.函数名

```
math.cos(1.5)
```

Out [4]: 0.0707372016677029

22.2 import 模块名 as 别名

In [5]:
> 第二种导入方法:
> import 模块名 as 别名
>
> 从原理上看,"import 模块名 as 别名"中的"别名"用户可以自行定义。但是,在数据分析和数据科学实践中,为了保证源代码的可读性,每个包或模块的"别名"并不是随意给出的,而是采用约定俗成的"别名",如 pandas、numpy 的"别名"一般为 pd 和 np。
>
> 在此方法中,函数的调用方法:
> 别名.函数名()

```
import math as mt
mt.sin(1.5)
```

Out [5]: 0.9974949866040544

22.3 from 模块名 import 函数名

In [6]:
> 第三种方法:
> from 模块名 import 函数名

22 模块函数

> **注意** 用此方法导入的模块中的函数的调用方法：
> 函数名()

```
from math import cos
cos(1.5)
```

> **提示** 建议与 In[3] 进行对比分析。此处解释器不报错的原因在于模块的导入方式发生了变化。

Out [6]: 0.0707372016677029

In [7]:
```
from math import sin
sin(1.5)
```

> **提示** 使用此方法直接从模块中导入指定函数，因此，在调用时只需使用函数名，而不需要另给出模块名，其调用方法与"内置函数"类似。

Out [7]: 0.9974949866040544

 扩展 为什么 Python 开发的代码感觉比 C/Java 慢一些？

在完成同一个功能时，为什么 Python 开发的代码感觉比 C/Java 慢一些？主要原因有 3 个：Python 是解释型语言；Python 采用的是 GIL（全局解释器锁）技术；Python 是一种动态语言。

如何编写高效率的 Python 代码？建议读者参阅 Luciano Ramalho 的书《Fluent Python: Clear, Concise, and Effective Programming》。

23 自定义函数

 常见疑问及解答。

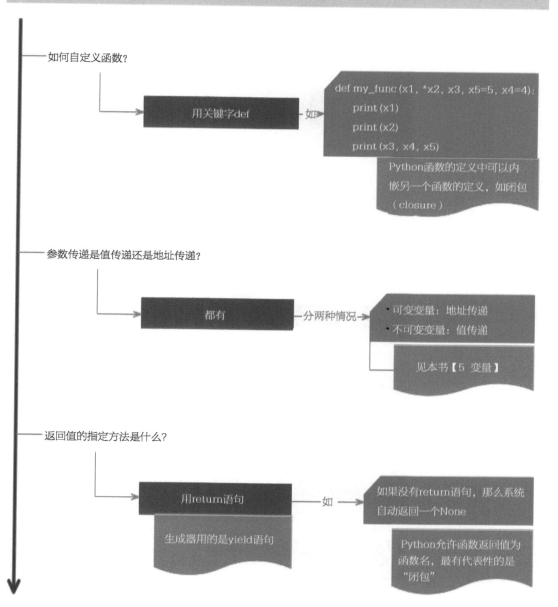

23 自定义函数

函数中的变量可见性如何定义或改变的？

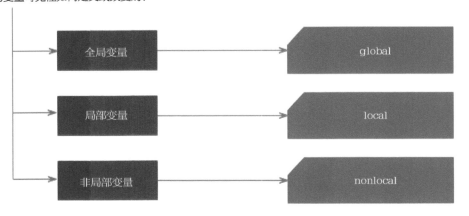

在Python中，函数编程应注意哪些问题？

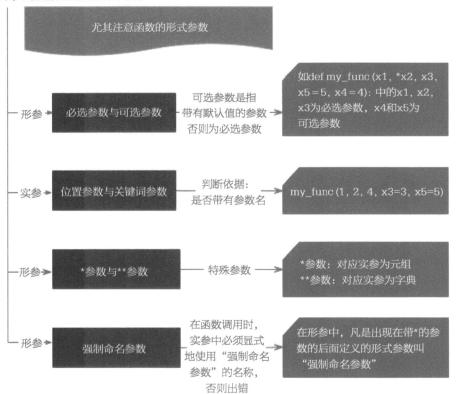

23.1 定义方法

In [1]:

> **提示** 与 C 和 Java 不同的是，Python 中的自定义函数是通过关键字 def 定义的。
>
> **注意** Python 中可以嵌套定义函数，即可以定义内嵌函数。如果内嵌函数 func2() 中调用了外围函数中的局部变量（而不是全局变量），则称为"闭包（closure）"。

```
def func1():
    j = 0
    print('hello world')
    def func2(i):
        print('pass'+str(i)+str(j))
    return func2
```

> **提示** 内嵌函数 func2() 为局部函数，只有在外围函数 func1() 内可见，也就是只能在函数 func1 内调用它。因此，外围函数中的 return func2 的作用是调用这个嵌套函数 func2()。如果没有 return func2，那么 func2() 是不会执行的，因为没有其他调用它的地方。

In [2]:

> **提示** 调用第一层函数的方法如下。

```
func1()
```

hello world

Out [2]: <function __main__.func.<locals>.func2>

In [3]:

> **提示** 调用第二层函数的方法如下。从 In[1] 的定义看，func1() 的返回值为 func2，因此，（从运行过程而言）可将此处的 func1()(2) 理解成 func2(2)。

```
func1()(2)
```

hello world

23 自定义函数

pass20

 提示 先运行 func1()，该函数 return func2 时运行了 func2()。如果没有这个 return 语句，那么系统会自动返回 None，导致的后果是 None(2)，错误提示为 'NoneType' object is not callable。

23.2 函数中的 docString

In [4]:

 提示 在定义函数时，可以（也建议）设置其 docString。

 注意 docString 部分需要用三个单引号括起来，此处三个单引号可以改为双引号。

```
def get_name(msg):
    '''根据用户提示 msg，获取用户名，如果输入为空，则默认为 Friend '''

    name = input(msg) or 'Anonymous User'
```

In [5]:

 提示 查看函数中的 docString 的方法——"help()" 函数或 "?"，参见本书【28 帮助文档（P181/In[1]）】。

```
help(get_name)
```
Help on function get_name in module __main__:

get_name(msg)

In [6]:

 提示 In[4]中代码的含义：根据用户提示 msg，获取用户名，如果输入为空，则默认为 Friend。

```
get_name?
```

23.3 调用方法

In [7]:

 提示 调用自定义函数——直接用函数名。

```
get_name('plz enter your name : ')
```

plz enter your name : chaolemen

Out [7]: 'chaolemen'

In [8]:

技巧 判断函数是否为"可被调用"的方法——内置函数 callable()。

```
print(callable(get_name))
```

Out [8]: True

23.4 返回值

In [9]:

提示 用 return 语句指定返回值。

```
def myfunc(i, j = 2):
    j = i+1
    return j
print(myfunc(3))
```

Out [9]: 4

In [10]:

提示 注意事项之一:如果没有 return 语句,那么函数返回值为 None。在 Python 中,通常用 None 表示缺失值。

```
def myfunc(i, j = 2):
    j = i+1
    #没有 return 语句,则返回 None
print(myfunc(3))
```

Out [10]: None

In [11]:

提示 注意事项之二:Python 函数可以同时返回多个值。

23 自定义函数

```
def myfunc(i, j = 2):
    j = i+1
    return i,j
```

> **提示**: return i,j 相当于 return (i,j)，见本书【15 元组（P101/In[22]）】。

```
a,b = myfunc(3)
a,b
```

Out [11]: (3, 4)

23.5 形参与实参

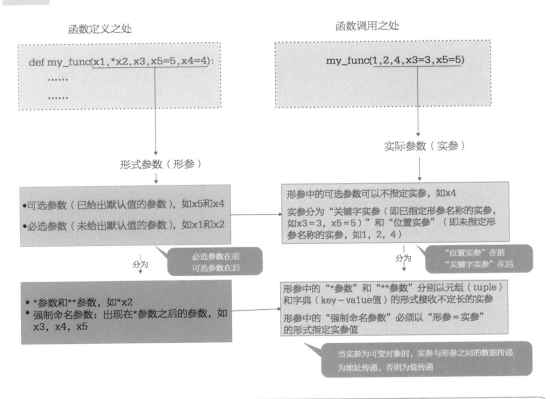

In [12]:
```
def my_func(x1, *x2, x3, x5 = 5, x4 = 4):
    print(x1)
```

```
print(x2)
print(x3)
print(x4)
print(x5)
```

 从函数定义视角看，形参分为可选和必选。

#判断依据：带有默认值的参数称为"可选参数"，调用它时可以不用给出实参
#x4 和 x5 为可选

```
my_func(1,2,4,x3 = 3,x5 = 5)
```

 从函数调用视角看，实参分为位置参数和关键词参数（又称为命名参数）

#判断依据是否带有参数名
#如 x3=3，x5=5 为关键字参数，1,2,4 为位置参数

 所有关键字参数必须出现在位置参数之后。

#否则报错：SyntaxError: positional argument follows keyword

Out [12]: 1
(2, 4)
3
4
5

In [13]:

 从函数定义视角看，在形参中，带"*"的参数接收的"实参"是元组。

#x2

 对号入座后，剩下的（2和4）变成一个元素传入参数 x2。

23 自定义函数

```
#(2, 4)
#本质：可接受可变数量的实参
```

> **提示**：从函数定义视角看，形参中带 "**" 的参数接收的是字典，即带有 key 的容器。

In [14]:
```
my_func(1,2,x4 = 4,x3 = 3,x5 = 5)
```

> **提示**：对应的 "函数定义" 的头部为：def my_func(x1,*x2,x3,x5 = 5, x4 = 4):

Out [14]:
```
1
(2,)
3
4
5
```

In [15]:

> **提示**：从函数定义视角看，在形参中，凡是出现在带 "*" 的参数的后面定义的形式参数叫 "强制命名参数"

```
#如 def my_func(x1,*x2,x3,x5 = 5,x4 = 4):中的 x3, x5 和 x4
```

> **注意**：在函数调用时，实参中必须显式地使用参数名，否则 Python 解释器会报错。

```
my_func(1,2,4,x3 = 3,x5 = 5)
```

Out [15]:
```
1
(2, 4)
3
4
5
```

23.6 变量的可见性

In [16]:

> **提示**：local 变量：在函数内定义的变量，仅在该函数内可见。

```
x = 0
def myFunc(i):
    x = i
    print(x)
```

 注意 第二个 x 与第一行中的 x 并非是同一个 x，第二个 x 是 local 变量。

```
myFunc(1)
print(x)
```

Out [16]: 1
0

In [17]:

 注意 local 改为 global 的方法：加一行 global x，而不是只写一个单词 global。

```
x = 0
def myFunc(i):
    global x
```

 提示 此行的意思为，接下来的 x 是 global 变量，而不是 local 变量。

```
    x = i
    print(x)

myFunc(1)
print(x)
```

 注意 global x 必须独占一行，且不能写成 global。

Out [17]: 1
1

In [18]:

 提示 与 global 类似，Python 中还有 nonlocal 变量，用法与 global 一样，但是用于内嵌函数。

23 自定义函数

```
x = 0
def myFunc(i):
    x = i
    def myF():
        nonlocal x        #这个必须是独立一行
        x = 2
        print(x)
    print(x)

myFunc(1)
print(x)
```

>
> **注意** 在输入源代码时，global x 与 nonlocal x 应以"独占一行"的格式缩写。

Out [18]:　1
　　　　　　0

23.7 值传递与地址传递

In [19]:
>
> **提示** 参数传递规则。
>
> #不可变对象（int、float、str、bool、tuple）：值传递，即形参的变化不会影响实参的值，原因是二者分别指向不同的内存空间。
> #可变对象（list、set、dict）：地址传递与上述"值传递"相反。

In [20]:
>
> **提示** （1）值传递：当实参为不可变对象（int、float、str、bool、tuple）时，实参和形参分别占用不同的内存空间，即在"被调用函数"中修改形参时，不会改变实参的值。
>
> ```
> i = 100
> def myfunc(j, k = 2):
>     j += 2
> myfunc(i)
> print(i)
> ```

注意 有默认值的参数在 Python 中是"可选参数"。

Out [20]: 100

In [21]:
 提示 （2）地址传递：当实参为可变对象（list、set、dict）时，实参和形参共享同一个内存空间，即当形参发生变化时，实参也会随之变化。

```
i = [100]
def myfunc(j, k = 2):
    j[0] += 2
myfunc(i)
print(i)
```

 注意 这个输出结果与 In[20]不同。

Out [21]: [102]

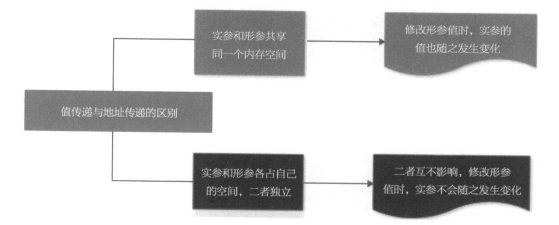

In [22]:
 提示 实参和形参的数据传递原则是对号入座，除了 self、cls 等特殊参数，这些特殊参数不需要另传给实参。

```
# def class_func(cls):
# def __init__(self, name, age):
```

23.8 其他注意事项

In [23]:

 在自定义函数时,初学者需要注意以下三个问题。

In [24]:

 第一,参数分为"位置参数"和"关键字参数"。

#区别在于是否带默认值,若带,则称为"关键字参数"

 关键字参数必须在位置参数之后,否则报错 SyntaxError: non-default argu ment follows default argument。也就是说,关键字参数后不能再出现非关键字参数。

```
def myfunc(j, k = 2):
    j += k
    j
```

 没有 return 语句,则返回 None。
在 Python 中,通常用 None 表示缺失值。

```
d = myfunc(2,3)
d
```

In [25]:
```
def myfunc(k = 2, j):
    j += k
    j
d = myfunc(2,3)
d
```

 报错 SyntaxError: non-default argument follows default argument。原因分析:关键字参数必须在位置参数后。

Out [25]: File "<ipython-input-25-9236e2f3931a>", line 1
    def myfunc(k=2, j):
                    ^
SyntaxError:non-default argument follows default argument

In [26]:
> 注意 　第二，如果不写 return 语句，那么返回值为 None。None 只有通过 print() 才能显示。

```
def myfunc(j, k = 2):
    j += k
    j
```

 提示 　若没有 return 语句，则自动返回 None。

```
d = myfunc(3)
print(d)
```

 提示 　None 的输出：若不用内置函数 print()，则无显示。

Out [26]: None

In [27]: `d is None`

Out [27]: True

In [28]:
> 提示 　第三，在 Python 中，函数也是对象，即"Python 认为一切皆为对象"。

```
myfunc = abs
print(type(myfunc))
```

 提示 　与其他对象一样，Python 函数名可作为 type() 的参数，返回值为函数类型。

```
print(myfunc(-100))
```

Out [28]: `<class 'builtin_function_or_method'>`
100

24 lambda 函数

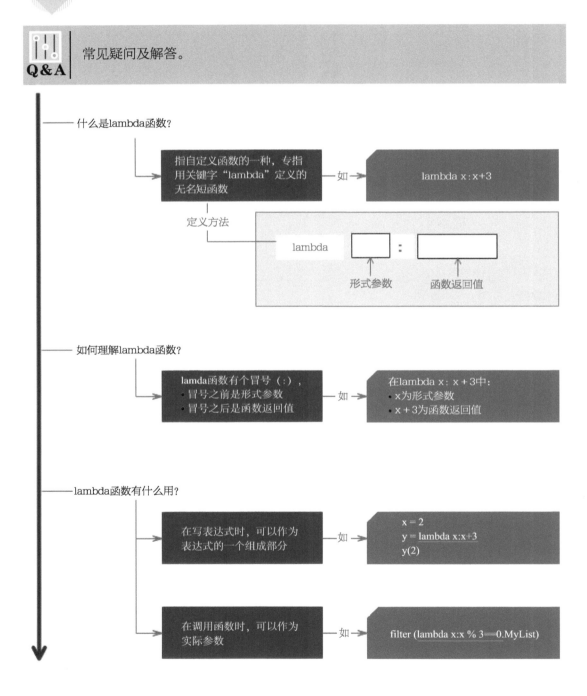

24.1 lambda 函数的定义方法

In [1]:

> **思路** lambda 函数的本质——单行匿名函数。

```
#冒号之后为返回值
#冒号之前为形式参数
x = 2
y = lambda x:x+3
y(2)
```

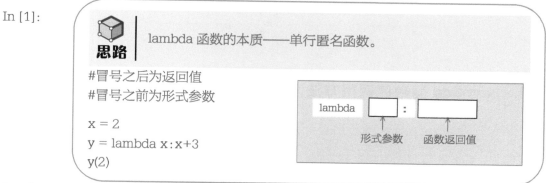

Out [1]:   5

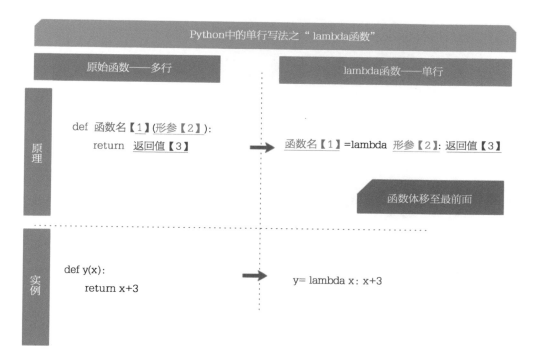

In [2]:

> **提示** In[1]中的 lambda 函数相当于如下普通函数。

24 lambda 函数

```
x = 2
def myfunc(x):
    return x+3
myfunc(2)
```

Out [2]: 5

24.2 lambda 函数的调用方法

In [3]:

提示 在数据分析与数据科学项目中，lambda 函数通常以另一个函数的参数形式使用，以 filter()函数为例。

```
MyList = [1,2,3,4,5,6,7,8,9,10]

filter(lambda x: x %3 == 0, MyList)
```

注意 filter 函数采用迭代式读取方式，从第二个参数（MyList）中按下标（位置）顺序依次读取每一个元素，并将其读入至第一个参数（lambda 函数）的 x。

Out [3]: <filter at 0x220666f6048>

In [4]:

提示 filter()函数的返回值为迭代器，需要进行强制类型转换才能看到其数据值，参见本书【25 迭代器与生成器（P165）】。通过强制类型转换方式显示 filter()函数的返回值。

Out [4]: [3, 6, 9]

In [5]:

提示 以 map()函数为例。

```
list(map(lambda x: x * 2, MyList))
```

Out [5]: [2, 4, 6, 8, 10, 12, 14, 16, 18, 20]

In [6]:

提示 以 reduce()函数为例。

```
from functools import reduce
reduce(lambda x, y: x + y, MyList)
```

 函数 map() 与 reduce() 的区别：reduce() 函数从 Python 3 版本开始不再是内置函数。

Out [6]:   55

 通往数据科学家和大数据分析师的捷径。

参与开源项目和竞赛是通往数据科学家和大数据分析师的两条捷径。建议初学者学会使用 GitHub 等开源协作社区/工具，提升自己的编程、协作和实战能力。KDnuggets 等机构每年都会公布 GitHub 顶级开源项目的排名和分析报告，值得初学者关注。相对于 2016 年的报告，2018 年《Top 20 Python AI and Machine Learning projects on Github》报告主要有如下几个变化。

（1）从贡献者（Contributors）的基数看，TensorFlow 已上升至排名第一；Scikit-learn 下降至第二，但其贡献者基数仍很大。

（2）从贡献者数量的增长率看，增长最快的项目分别为：TensorFlow（169%）、Deap（86%）、Chainer（83%）、Gensim（81%）、Neon（66%）、Nilearn（50%）。

（3）2018 年的新项目：Keras（贡献者数量：629）和 PyTorch（贡献者数量：399）。

GitHub 官网及本书作者主页如图 24-1 所示。

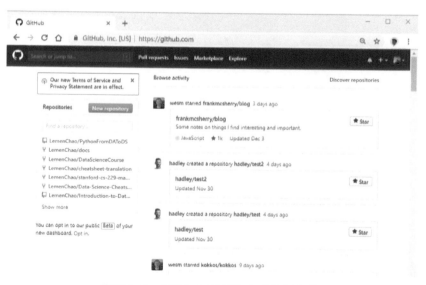

图 24-1　GitHub 官网及本书作者主页

第三篇
Python 进阶

- 迭代器与生成器
- 模块
- 包
- 帮助文档
- 异常与错误
- 程序调试方法
- 面向对象编程
- 魔术命令
- 搜索路径
- 当前工作目录

25 迭代器与生成器

 常见疑问及解答。

— 为什么Python中不显示有些函数的返回值，如range()？
  → 该函数的返回值是迭代器（iterator） → 迭代器在Python中的应用最为广泛

  L. Peter Deutsch 曾说 "To iterate (迭代) is human, to recurse (递归) divine."

— 什么是迭代器？
  → 可被next()函数调用，并不断返回下一个值的对象 → Python内置函数iter()可以将"可迭代对象"转换成"迭代器"

— 如何遍历迭代器？
  → 内置函数next() → 也可以用在for语句中

— 什么是生成器？
  → 指生成一个新的迭代器的函数 —特点→
    • 不用return语句，而用yield语句
    • 不是立刻计算，而是惰性计算

— 如何定义生成器？
  → 与普通函数定义类似，但需要将return改为yield —如→
  ```
  def myGen():
      x=range(1,11)
      for i in x:
          yield i+2
  ```

25 迭代器与生成器

25.1 可迭代对象与迭代器

In [1]:

> **注意** 在 Python 中，"可迭代对象"和"迭代器"是两个既有区别又有联系的概念。

#（1）能够接收"可迭代对象"的函数都可以接收"迭代器"
#（2）用 Python 内置函数 iter() 可以将"可迭代对象"转换成"迭代器"

> **提示** 可迭代对象（iterable object）：可以直接作用于循环语句（如 for 语句）的对象。

> **提示** 迭代器（iterator）：可以被内置函数 next() 调用，并不断返回下一个值的对象。

In [2]:

> **提示** （1）可迭代对象不一定是迭代器。

```
myList = [1,2,3,4,5]
next(myList)
```

> **提示** 报错，TypeError: 'list' object is not an iterator。原因分析：myList 虽属于"可迭代对象"，但不是"迭代器"。

> **提示** 判断方法：用内置函数 ininstance() 和 collections 模块。

```
from collections import Iterable
isinstance (myList, Iterable)
```

Out [2]:
```
---------------------------------
TypeError                                 Traceback (most recent call last)
<ipython-input-2-4b84f2f2bfca> in <module>()
      1 myList=[1,2,3,4,5]
----> 2 next(myList)
```

In [3]:

> **提示** （2）将可迭代对象转换为迭代器——内置函数 iter()。

```
myIterator = iter(myList)
print(next(myIterator))
print(next(myIterator))
print(next(myIterator))
```

> 提示 遍历迭代器内容方法——内置函数 next()。

Out [3]:
```
1
2
3
```

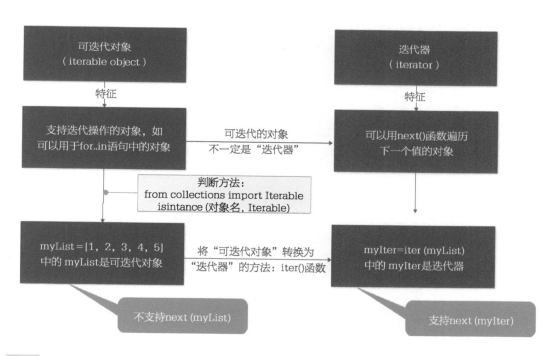

25.2 生成器与迭代器

In [4]:

> 提示 生成器（generator）是指生成一个新的迭代器的函数，与普通函数的区别如下。

25 迭代器与生成器

In [5]:
```
#第一，不用 return 语句，而用 yeild 语句。
#第二，不是"立刻计算"，而是"惰性计算"——在调用生成器时，不会
立刻执行它，而是推迟至需要调用其中的每个元素时才运行。
```

（1）生成器的定义方法与一般函数不同，用的是 yeild 语句，而不是 return 语句。

```
def myGen():
    x = range(1,11)
    for i in x:
        yield i+2
```

In [6]:

（2）生成器的特点：生成器不会被立即返回结果，因为生成器遵循的是"惰性计算"模式。

```
myGen()
```

输出结果为<generator object myGen at 0x000002442F11A8E0>，而不是具体返回值。

Out [6]: <generator object myGen at 0x000002442F11A8E0>

In [7]:

（3）生成器的特点：只有其中的元素被调用时才会被执行。

```
for x in myGen():
    print(x,end=",")
```

可以采用"print()+*+生成器"进行直接显示生成器内容，如 print(*myGen())

Out [7]: 3,4,5,6,7,8,9,10,11,12,

惰性计算技术。

本章知识中涉及了"惰性计算（lazy evaluation）"技术，有时被称为"惰性求值"技术。惰性计算技术是大数据分析和数据科学项目中常用的技术之一，也是 Spark 的关键技术之一，参见本书【46.Spark 编程（P433/In[10]）】中的惰性计算技术。

26 模块

 常见疑问及解答。

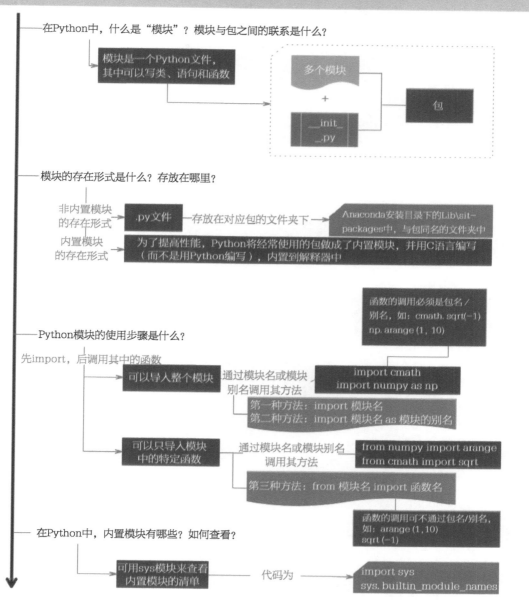

26 模块

26.1 导入与调用用法

In [1]:
 第一种方法:
import 模块名

在此方法下,调用函数的方式为:
模块名.函数名()

```
import math
math.sin(1.5)
```

Out [1]: 0.9974949866040544

In [2]:
```
cos(1.5)
```

 报错:NameError: name 'cos' is not defined。
原因分析:调用方法错误,缺少模块名 math。

Out [2]:
```
---------------------------------
NameError                                 Traceback (most recent call last)
<ipython-input-2-25c1f7923f0b> in <module>()
---> 1 cos(1.5) #报错,NameError: name 'cos' is not defined

NameError: name 'cos' is not defined
```

In [3]:
 第二种方法:
import 模块名 as 别名

 在此方法下,调用函数的方式为:
别名.函数名()

```
import math as mt
mt.sin(1.5)
```

Out [3]: 0.9974949866040544

In [4]:
 第三种方法:
from 模块名 import 函数名

> **注意** 在此方法下，调用函数的方法为：
> 函数名()
>
> ```
> from math import cos
> cos(1.5)
> ```

Out [4]: 0.0707372016677029

26.2 查看内置模块清单的方法

In [5]:
> **提示** 查看内置模块清单的方法：可用 sys 模块来查看内置模块的清单。内置模块不需要 pip install 或 conda install，直接 import 即可。
>
> ```
> import sys
> sys.builtin_module_names
> ```

Out [5]: ('_ast',
'_bisect',
'_blake2',
'_codecs',
'_codecs_cn',
'_codecs_hk',
'_codecs_iso2022',
'_codecs_jp',
'_codecs_kr',
'_codecs_tw',
'_collections',
'_csv',
'_datetime',
'_findvs',
'_functools',
'_heapq',
'_imp',
'_io',
'_json',
'_locale',

26 模块

```
'_lsprof',
'_md5',
'_multibytecodec',
'_opcode',
'_operator',
'_pickle',
'_random',
'_sha1',
'_sha256',
'_sha3',
'_sha512',
'_signal',
'_sre',
'_stat',
'_string',
'_struct',
'_symtable',
'_thread',
'_tracemalloc',
'_warnings',
'_weakref',
'_winapi',
'array',
'atexit',
'audioop',
'binascii',
'builtins',
'cmath',
'errno',
'faulthandler',
'gc',
'itertools',
'marshal',
'math',
'mmap',
'msvcrt',
'nt',
'parser',
```

```
'sys',
'time',
'winreg',
'xxsubtype',
'zipimport',
'zlib')
```

  扩展 | Python 语言与 R 语言的纠葛。

近年来，一方面，民间对 Python 与 R 语言的对比分析"如火如荼"，给人感觉是"你死我活"；另一方面，业界开始探讨 Python 和 R 代码的集成应用，好像是"你中有我，我中有你"。

最具代表性的是 Wickham（R 领袖）和 McKinney（Python 领袖）都在主张这两种语言的融合式发展，如图 26-1 所示。用 Wickham 的话讲，未来不是"Python vs R"，而是"Python and R"。例如，2018 年，McKinney 的 Ursa Labs 宣布，即将与 Wickham 的 RStudio 合作，致力于改进 R 和 Python 语言本身及其用户体验。

就目前而言，我们有没有可能集成运用（或交叉应用）Python 和 R？答案是"可以"，主要途径有以下三个。

（1）在 Python 代码中调用 R 代码——用面向 R 的 Python 包，如 rpy2、pyRserve、Rpython（rpy2 扩展使下面的 Jupyter 成为可能）等。

（2）在 R 代码中调用 Python 代码——用面向 Python 的 R 包，如 rPython、PythonInR、reticulate、rJython、SnakeCharmR、XRPython。

（3）用 Jupyter Notebook/Lab——安装两种语言的内核即可。

图 26-1　Wickham、Sandy Ryza 和 McKinney（从左到右，来源：Twitter）

27 包

 常见疑问及解答。

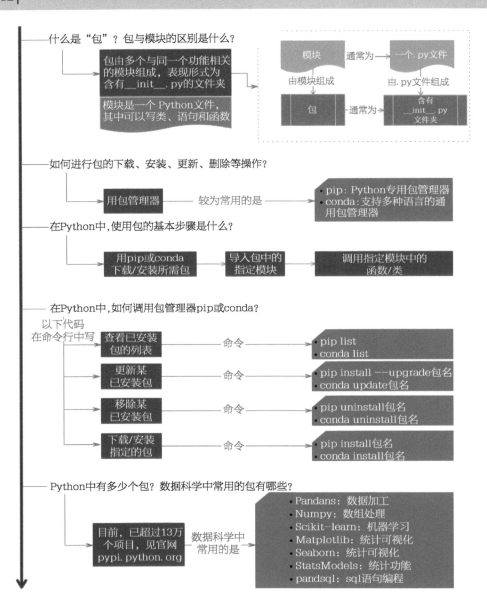

27.1 相关术语

In [1]:

> **提示** Python 中的文件、包、模块、库等术语的区别与联系。

```
#文件：物理上的组织方式，如一个 math.py 的程序文件可以作为
 module 的文件类型有".py"".pyo"".pyc"".pyd"".so"".dll"
#包：通常为一个文件夹，即含有__init__.py 文件的文件夹
#模块：逻辑上的组织方式，如 math
#库：模块的集合
```

> **提示** 常用的 Python 包管理器（包的下载、更新、删除、查看等）有两种，二者的区别如下。

```
#PIP 工具：仅管理 Python 的包，对应的包服务器为 PyPI（Python
 Packages Index），URL 为 https://pypi.org/
#Conda 工具：管理多种语言的包，对应的包服务器为 Conda，URL
 为 https://conda.io
```

27.2 安装包

In [2]:

> **提示** 用 PIP 工具或 Conda 工具，用法：分别在命令行（Anaconda Prompt）中输入如下命令。

```
#pip install 包名
#conda install 包名
```

27 包

 当用 PIP 下载某个包很慢时，可以考虑用 Conda 试试。

 为了方便编程，在 Jupyter Notebook 中已经预安装了数据分析和数据科学中的常用包，所以本书中包的导入之前一般不需要下载和安装，除了【46 Spark 编程（P425）】中的 pspark 等特殊包。

如上图中，输入pip install scipy时提示Requirement already satisfied: scipy in c:\anaconda\lib\sitepackages，说明此包已安装，不需要另行安装。但是pip install orderPy无此提示。

27.3 查看已安装包

In [3]:

 用 PIP 工具或 Conda 工具，用法为：在命令行（Anaconda Prompt）中输入如下命令。

pip list
conda list

27.4 更新（或删除）已安装包

In [4]:

 用 PIP 工具或 Conda 工具，用法：分别在命令行（Anaconda Prompt）中输入如下命令。

```
#pip install --upgrade 包名
#conda update 包名
```

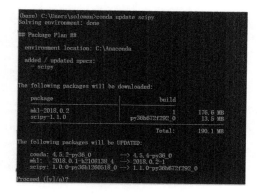

> **提示** 删除已安装包的命令为 pip uninstall 和 conda uninstall。

27.5 导入包

In [5]:

> **注意** 与 In[2]~In[4]不同的是，以下代码在 Jupyter notebook 中输入，参见本书【26 模块（P170/In[3]）】。

> **提示** （1）导入一个模块：建议在 as 后面尽量用约定俗成的别称。

```
import pandas as pd
```

> **提示** （2）一次导入多个模块：用逗号分隔。

```
import pandas as pd, numpy as np, math as math
```

> **注意** 不要写成：import pandas,numpy,math as pd,np,ma

27 包

 （3）只导入一个模块中的特定函数。

from pandas import DataFrame

 在"from pandas import DataFrame"中，第一个参数 pandas 为模块名，第二个参数 DataFrame 为这个模块中的一个函数名。

 from 后面是模块名，import 后面是函数名。

 （4）有层次的文件结构的模块的导入，用句点表示层次关系。

import Graphics.Primitive.fill

 如果用 pip 和 conda 命令无法下载或安装某包，还可以到该包的官网下载，并按官网操作提示直接安装即可。

27.6 查看包的帮助

In [6]:

 在数据分析和数据科学项目中，可以通过很多方法查看某个包的帮助，建议初学者参考如下。

```
# Jupyter Notebook 中的 help 菜单中提供了常用包的官网地址，如
  Pandas、NumPy 等。
#查阅特定包的官网，如 NumPy 的官网说明文档的 URL：https://
  docs.scipy.org/doc/numpy/reference/?v=20180107143112
#用包提供的内置属性和方法，如查看包的版本号
  pd.__version__
```

 version 的前后各有 2 个下画线。

Out [6]:   '0.22.0'

27.7 常用包

In [7]:

> **提示** 在数据分析和数据科学项目中，常用的基础包包括如下几种。
>
> #Pandas：数据框（关系表）和 Series 处理
> #Numpy：多维数组（矩阵）处理
> #Scikit-learn、TensorFlow 和 PyTorch：机器学习
> #Matplotlib：统计可视化
> #Seaborn：数据可视化
> #StatsModels：统计分析
> #pandsql：SQL 编程
> #Scrapy：Web 爬取
> #pySpark：Spark 编程
> #NLTK、spaCy：自然语言处理（英文）
> #pynlpir、Jieba：自然语言处理（中文）
> #wordclound：生成词云
> #random：生成随机数
>
> **提示** 更多内容参见本书【50.3 常用模块与工具包（P495）】和【50.5 核心机器学习算法（P496）】

 扩展 Jupyter Notebook 中的 Markdown。

在数据科学项目中，我们用 Word、Excel 还是 PowerPoint？其实，比 Word、Excel 和 PowerPoint 更专业的沟通工具是 Markdown。以 Jupyter Notebook 中的 Markdown 功能为例，主要特点如下。

（1）支持将源代码、注释、文字、段落、图表混排在一起。

（2）提供 PDF、Word、PPT、HTML 等多种导出形式。

（3）以代码块为单位，可以将执行结果直接插入到对应的程序代码块，进而方便数据科学家直接阅读程序源代码，不需要像程序员那样编译/执行程序代码，使自己的精力集中在"基于数据的管理问题"。

建议读者学会用 Jupyter Notebook 的 Markdown 功能。

28 帮助文档

 常见疑问及解答。

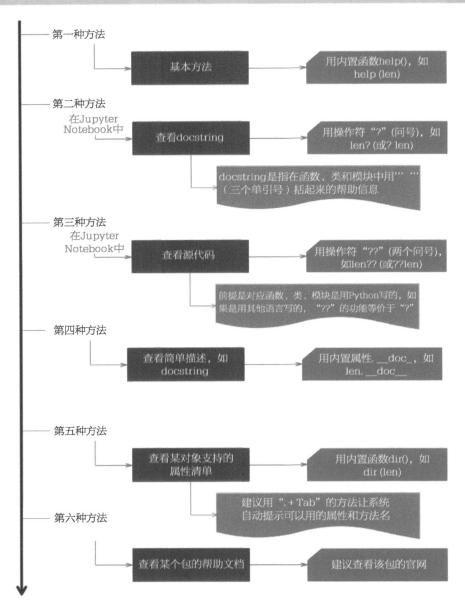

28.1 help 函数

In [1]:
> **提示** 最基本、最通用方法——用内置函数 help()。

```
help(len)
```

Out [1]: Help on built-in function len in module builtins:

len(obj, /)
Return the number of items in a container.

28.2 docString

In [2]:
> **提示** iPython 中的特殊方法——用 "?"。

```
len?
```

> **提示** iPython（或基于 iPython 的 Jupyter Notebook/Lab 等）系统显示的帮助信息如下。

#Signature: len(obj, /)
#docString: Return the number of items in a container.
#Type: builtin_function_or_method

> **注意** 此方法是 iPython 提供的功能，不是 Python 的语法。

In [3]:
```
myList1 = [1,2,3,4]
myList1.append?
```

> **提示** 系统显示的帮助信息如下。

#docString: L.append(object) -> None -- append object to end
#Type: builtin_function_or_method

28 帮助文档

In [4]:

思路 docstring 查阅的帮助文档为目标程序中用三个"""括起来的多行说明性文字。

```
def testDocString():
    """此处为 docString,
    即用"?"能查看得到的帮助信息"""
    return(1)

testDocString?
```

提示 iPython（或基于 iPython 的 Jupyter Notebook/Lab 等）系统显示的帮助信息如下。

```
#Signature: testDocString()
#docString:
#此处为 docString,即用"?"能查看得到的帮助信息
#File: C:\Users\Soloman\Clm\<ipython-input-4-742d4bf944c4>
#Type: function
```

In [5]:

?testDocString

提示 iPython（或基于 iPython 的 Jupyter Notebook/Lab 等）系统显示的帮助信息与 In[4]一致。

技巧 "?"放在最前面也可以。

28.3 查看源代码

In [6]:

提示 查看源代码的方法——用"??"。

testDocString??

提示 系统显示的帮助信息与In[4]一致。

注意 前提：目标对象是用Python编写的，否则无法查看源代码，"??"的功能变得与"?"一样。

In [7]:

提示 查看内置函数len()的帮助信息。

len?

提示 iPython（或基于iPython的Jupyter Notebook/Lab等）系统显示的帮助信息如下。

#Signature: len(obj, /)
#docString: Return the number of items in a container.
#Type: builtin_function_or_method

In [8]: len??

提示 "len??"与"len?"的输出结果一样。原因分析：内置函数len()不是用Python编写的。

28.4 doc 属性

In [9]:

提示 __doc__属性的前后各有两个下画线，是Python面向对象编程方法中为每个类自动增加的默认属性，参见本书【31 面向对象编程】。

testDocString.__doc__

Out [9]: '此处为docString，\n 即用"?"能查看得到的帮助信息'

In [10]: len.__doc__

Out [10]: 'Return the number of items in a container.'

28.5 dir()函数

In [11]:

> **提示** 查看某对象支持的方法或属性清单——内置函数 dir()。

```
dir(print)
```

Out [11]: ['__call__',
          '__class__',
          '__delattr__',
          '__dir__',
          '__doc__',
          '__eq__',
          '__format__',
          '__ge__',
          '__getattribute__',
          '__gt__',
          '__hash__',
          '__init__',
          '__init_subclass__',
          '__le__',
          '__lt__',
          '__module__',
          '__name__',
          '__ne__',
          '__new__',
          '__qualname__',
          '__reduce__',
          '__reduce_ex__',
          '__repr__',
          '__self__',
          '__setattr__',
          '__sizeof__',
          '__str__',
          '__subclasshook__',
          '__text_signature__']

In [12]:
```python
#再如
dir(len)
```

Out[12]: ['__call__',
'__class__',
'__delattr__',
'__dir__',
'__doc__',
'__eq__',
'__format__',
'__ge__',
'__getattribute__',
'__gt__',
'__hash__',
'__init__',
'__init_subclass__',
'__le__',
'__lt__',
'__module__',
'__name__',
'__ne__',
'__new__',
'__qualname__',
'__reduce__',
'__reduce_ex__',
'__repr__',
'__self__',
'__setattr__',
'__sizeof__',
'__str__',
'__subclasshook__',
'__text_signature__']

In [13]:
```python
dir?
```

 提示　Python 采用"鸭子类型编程（Duck Typing）"技术。"鸭子类型编程技术"是指不管是否为真的鸭子（即所属 class），只要看像不像鸭子（即支持哪些属性）和走起路、叫起来像不像鸭子（即支持哪些方法），如果是，就认为它是"鸭子"。

28.6 其他方法

In [14]:

 与 Ecipse、MyEclipse、Visual Studio 类似，程序员可以在 Jupyter Notebook 中用 Tab 或通配符 "*" 的方法进行自动提示和自动填充。

myList1.c*?

 Jupyter Notebook/Lab 显示的自动提示信息如下。

#myList1.clear
#myList1.copy
#myList1.count

 如何找到免费大数据？

除了利用网络爬虫收集数据、数据生成和存储部门的供给，我们还可以通过以下方式获得大数据。

（1）Googe Dataset Search 引擎：https://toolbox.google.com/datasetsearch。

（2）政府开放数据，如美国政府开放的数据集（https://www.data.gov/）等。

（3）企业或公益组织公布的数据，如 Amazon Web Services（AWS） datasets（https://aws.amazon.com/datasets/）等。

（4）大数据竞赛平台提供的数据，如 Kaggle 数据集（https://www.kaggle.com/datasets）。

（5）机器学习和统计学领域的经典数据集，如 UCI 数据集（https://archive.ics.uci.edu/ml/datasets.htm）。

29　异常与错误

Q&A 常见疑问及解答。

── iPython中如何更改异常信息的显示方式？
　　→ 用魔术命令 %xmode

── Python异常处理的模板是什么？

```
try:
    可能发生异常的语句

except Ex1:
    发生异常Ex1时要执行的语句

except (Ex2, Ex3):
    发生异常Ex2或Ex3时要执行的语句

except:
    发生其他异常时要执行的语句

else:
    无异常时要执行的语句

finally:
    不管是否发生异常，都要执行的语句
    如文件、数据库、图形句柄资源的释放
```

── Python中有哪些常见异常、错误？

```
• Exception          常规错误的基类
• AttributeError     对象没有这个属性
• EOFError           没有内建输入，到达EOF 标记
• IOError            输入/输出操作失败
• ImportError        导入模块/对象失败
• IndexError         序列中没有此索引（index）
• KeyError           映射中没有这个键
• MemoryError        内存溢出错误
• NameError          未声明/初始化对象（没有属性）
• NotImplementedError  尚未实现的方法
• SyntaxError Python  语法错误
• IndentationError   缩进错误
• TypeError          对类型有误
• ValueError         传入无效的参数
• Warning            警告的基类
• Swntaxwarning      可疑的语法警告
```

29 异常与错误

29.1 try/except/finally

In [1]:

注意 try、except、finally 之后都有 ":"。

```
try:
    f=open('myfile.txt','w')
    while True:
        x=input("请输入一个整数，若需要停止运行请输入字母'Q'")
        if x.upper()=='Q':break
        y=100/int(x)
        f.write(str(y)+'\n')
except ZeroDivisionError:
    print("抛出 ZeroDivisionError")
except ValueError:
    print("抛出 ValueError:")
finally:
    f.close()
```

思路 finally 部分：指不管有没有异常，都会运行的语句。可见，Python 中的异常处理除了进行异常处理，还可以进行句柄资源的释放。除了 try 语句，Python 中的 with 语句也支持异常处理和句柄资源的释放。

注意 与 C 和 Java 不同的是，Python 的异样处理中还可以加 else 语句——不发生任何异常时运行的语句。

注意 finally 与 else 的区别。

Out [1]: 请输入 Qa
请输入 QQ

> **提示** Python 中 try/except/finally 语句的语法模板如下：

```
try:
    可能发生异常的语句
except Ex1:
    发生异常Ex1时，要执行的语句
except (Ex2, Ex3):
    发生异常Ex2或x3时，要执行的语句
except:
    发生其他异常时，要执行的语句
else:
    无异常时，要执行的语句
finally:
    不管是否发生异常，都要执行的语句，如文件、数据库、图形句柄资源的释放
```

29.2 异常信息的显示模式

In [2]: 提示 | 在 Python 中异常信息的显示模式（%xmode）有三种，即 Context（默认值）、Plain 和 Verbose。

In [3]: 提示 | Plain 模式如下。

```
%xmode Plain
x = 1
x1
```

Out [3]: Exception reporting mode: Plain
Traceback (most recent call last):

   File "<ipython-input-14-6067cb69f3f6>", line 3, in <module>
     x1

NameError: name 'x1' is not defined

In [4]: 提示 | Verbose 模式如下。

```
%xmode Verbose
x = 1
x1
```

 思路 | Python 中定义了很多异常（Exceptions）和错误（Errors）类，参见 Python 官网，如 https://docs.python.org/2/tutorial/errors.html。

Out [4]: Exception reporting mode: Verbose

NameError   Traceback (most recent call last)
<ipython-input-15-9ace4a24824c>in <module>()
    1 get_ipython().magic('xmode Verbose')
    2 x=1

29 异常与错误

```
----> 3 x1
    globalx1 = undefined

NameError: name 'x1' is not defined
```

In [5]:

 提示 | Context 模式如下。

```
%xmode Context
x = 1
x1
```

Out [5]:
```
Exception reporting mode: Context
------------------------------------
NameError                  Traceback (most recent call last)
<ipython-input-12-5d9d36673da7>in <module>()
  1 get_ipython().magic('xmode Context')
  2 x = 1
----> 3 x1

NameError: name 'x1' is not defined
```

29.3 断言

In [6]:

 提示 | 在数据分析、数据科学中，断言（Assertion）主要用于"设置检查点（Check Points）"。

 注意 | assert 之后为检查条件，当此条件为假时，抛出断言。

```
a = 1
b = 2
assert b != 0,"分母不能等于0"
```

 提示 | 条件为真，则不抛出 AssertionError。

> **注意** 别忘了 assert 语句中的逗号。

In [7]:
```
a = 1
b = 0
assert b != 0,"分母不能等于 0"
```

> 👁 **提示** 条件为假，则抛出 AssertionError。

> 👁 **提示** 输出结果为 Assertion Error，具体如下：

```
#AssertionError: 分母不能等于 0
```

Out [8]:
```
---------------------------------------------
AssertionError          Traceback (most recent call last)
<ipython-input-14-f99003c3453f>in <module>()
  1 a=1
  2 b=0
----> 3 assertb!=0,"分母不能等于 0"
  4

AssertionError: 分母不能等于 0
```

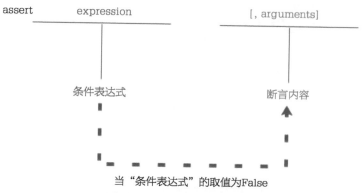

30 程序调试方法

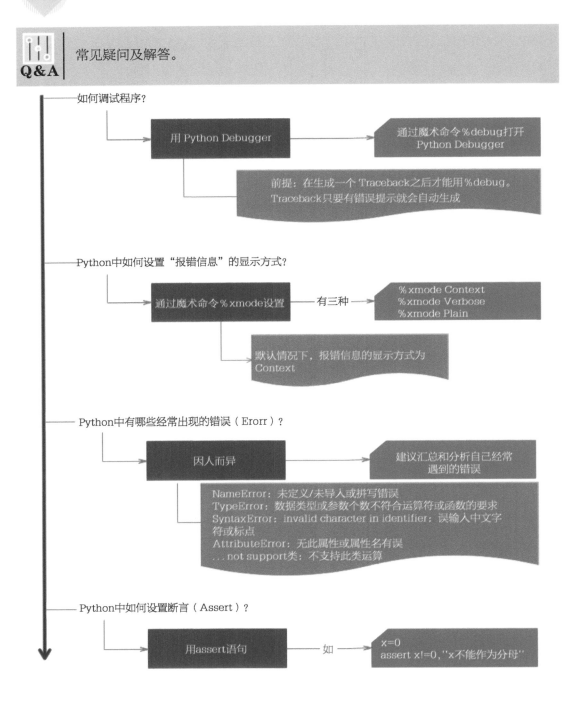

30.1 调试程序的基本方法

In [1]:
 思路   当 Python 抛出异常或错误信息时，可以用 PDB（Python Debugger, Python 调试器）进行程序调试。

x = 1
x1

 提示   报错：
NameError: name 'x1' is not defined

Out [1]: ---------------------------------------

NameError   Traceback (most recent call last)
<ipython-input-1-b36d888da5d8>in <module>()
    1 #【提示】当 Python 抛出异常或错误信息时，可以用 PDB（Python Debugger，Python 调试器）进行调试程序。
    2 x = 1
----> 3 x1

NameError: name 'x1' is not defined

In [2]:
 提示   打开 PDB 的方法是输入魔术命令：
%debug

%debug

 注意   退出 PDB 的方法：在 PDB 中输入命令 q 或 quit。

 提示   除了 PDB，常用的 Python 调试器有 Pylint、Pychecker 等。

Out [2]: > <ipython-input-1-b36d888da5d8>(3)<module>()
    1 x = 1
----> 2 x1

ipdb> x

30 程序调试方法

```
1
ipdb> x1
*** NameError: name 'x1' is not defined
ipdb> x
1
ipdb> q
```

30.2 设置错误信息的显示方式

In [3]:
> 提示　在 Python 中，异常信息的显示模式（%xmode）有三种，即 Context（默认）、Verbose 和 Plain，参见本书【29 异常与错误】。
>
> ```
> %xmode Plain
> y = 1
> Y
> ```

Out [3]: Exception reporting mode: Plain
Traceback (most recent call last):

　　File "<ipython-input-3-2ab22ce97508>", line 4, in <module>
　　　Y

NameError: name 'Y' is not defined

In [4]:
```
%xmode Verbose
y = 1
Y
```

Out [4]: Exception reporting mode: Verbose

NameError   Traceback (most recent call last)
<ipython-input-4-4e5fe2f91061>in <module>()
　1 get_ipython().run_line_magic('xmode', 'Verbose')
　2 y = 1
----> 3 Y
　　global Y = undefined

NameError: name 'Y' is not defined

In [5]: %debug

Out [5]: > <ipython-input-4-4e5fe2f91061>(3)<module>()
              1 get_ipython().run_line_magic('xmode','Verbose')
              2 y = 1
        ----> 3 Y

        ipdb> y
        1
        ipdb> Y
        *** NameError: name 'Y' is not defined
        ipdb> y
        1
        ipdb> quit

30.3 设置断言的方法

In [6]:
 关于断言，参见本书【29 异常与错误（P190/In[6]）】。

a = 1
b = 0
assert b != 0,"分母不能等于0"

 条件为假，则抛出 AssertionError。

 在编写代码时，别忘了 assert 语句中的逗号。

Out [6]: ---------------------------------------
         AssertionError   Traceback (most recent call last)
         <ipython-input-6-ada778478a55>in <module>()
             2 a = 1
             3 b = 0
         ----> 4 assert b != 0,"分母不能等于0"
             5

30 程序调试方法

```
          6 #提示：条件为假，则抛出 AssertionError

AssertionError: 分母不能等于 0
```

In [7]: `%debug`

Out [7]:
```
> <ipython-input-6-ada778478a55>(4)<module>()
      2 a = 1
      3 b = 0
----> 4 assert b != 0,"分母不能等于 0"
      5
      6 #提示：条件为假，则抛出 AssertionError

ipdb> a
ipdb> b
ipdb> a
ipdb> quit
```

扩展 ｜ 一图看懂 Python 语法，如图 30-1 所示。

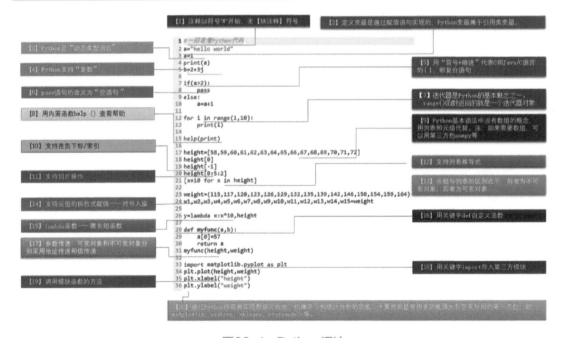

图30-1　Python语法

31 面向对象编程

 常见疑问及解答。

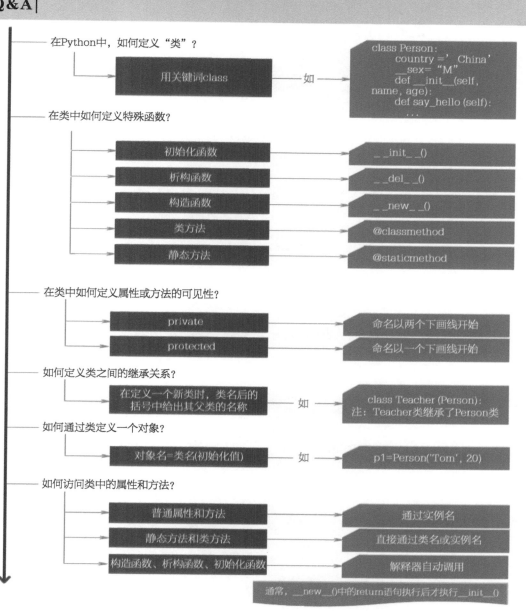

31 面向对象编程

31.1 类的定义方法

In [1]:

提示 | Python 面向对象编程的基本思路与 C++、C#和 Java 是一致的——遵循面向对象思想,因此本书不介绍面向对象思想本身,主要讲解 Python 的特殊之处。

思路 | 在 Python 中,定义一个类主要用"关键字 class +类名+冒号+缩进"方式实现。

```
class Person:
    nationality = 'China' #此处将国籍/nationality定义为public属性
    _deposit=10e10 #Python受保护属性名以一个下画线开始,此处将存
                    款数量/deposit定义为protected属性
    __gender="M" #Python私有属性名以两个下画线开始,此处将性别
                    /gender定义为private属性

    def __init__(self, name, age):
        self.name = name #实例属性
        age = age #局部变量

    def say_hi(self):
        print(self.name)

p1 = Person('Tom', 30)
p1.say_hi()
```

注意 | Python 中属性和方法的可见性的定义方法与 C++、C#和 Java 不同,属性/方法名中分别以一个下画线和两个下画线为开始来区分受保护类型(protected)和私有类型(private)。

提示 | Person('Tom', 30) 做了两件事:新建(new)一个新对象,并对此进行初始化(init)后得到一个实例(p1)。

提示 | Python 中的三个重要函数如下。函数名的前后各有两个下画线。

#__init__():初始化函数

```
#__new__(): 构造函数
#__del__(): 析构函数
```

Out [1]:   Tom

31.2 类中的特殊方法

In [2]:

 提示 除了__init__()、__new__()和__del__(), Python中的常用方法可以分为三种: 实例方法、类方法和静态方法。

```python
class Person:
    """
    此处为类Person的docString
    """
    nationality = 'China'
    _deposit=10e10
    __gender="M"

    def __init__(self, name, age):
        age = age #age为函数__init__()中的局部变量
        self.name = name #与age不同的是，self.name为实例属性
```

 提示 实例方法的定义语法与一般函数的定义语法类似, 区别在于形参。

```python
    #实例方法的第一个参数必须为位置参数self, 否则提示: TypeError:
     ***() 函数的positional arguments 错误。
    #self的含义
       #当前实例的一个引用, 表示该方法为"实例方法"
       #实例方法可以通过"实例名.函数名"的形式访问

    def say_hi(self):
        print(self.name)
```

 提示 类方法的定义方法: 在该方法的前面加一行。
@classmethod

31 面向对象编程

注意 | 在类方法的定义中,第一个参数必须为"类的引用 cls",即可以通过类名调用该函数。

```
@classmethod
def class_func(cls):
    cls.nationality = 'CHINA'
    print('I live in {0}'.format(cls.nationality))
```

提示 | 静态方法的定义方法:在该方法的前面加一行。
@staticmethod

```
#静态方法的特点:形参中没有 cls 或 self,甚至可以没有任何参数
```

思路 | 没有任何参数的函数一般定义为"静态方法"。

```
@staticmethod
def static_func(x, y):
    print(x+y)

p1 = Person('Tom', 20)
p1.say_hi()
```

Out [2]: Tom

In [3]:

提示 | "静态方法"的访问方法:通过类名和实例名均可。

```
Person.static_func(200,300)    #通过类名调用"静态方法"
```

Out [3]: 500

In [4]:
```
p1.static_func(200,300)        #通过实例名调用"静态方法"
```

Out [4]: 500

In [5]:

提示 | "类方法"的访问方法:通过类名和实例名均可。

```
Person.class_func()            #通过类名调用"类方法"
```

Out [5]: I live in CHINA

In [6]: p1.class_func()   #通过实例名调用"类方法"

Out [6]: I live in CHINA

31.3 类之间的继承关系

In [7]:
 提示　Python 中表示"类"之间的"继承"关系的方式比较特殊：在定义一个类（如 Teacher）时，将其父类名（如 Person）放在该类名之后的括号中。

```
class Teacher(Person):
    pass
t1 = Teacher("zhang",20)
```

In [8]: Person.class_func()

Out [8]: I live in CHINA

In [9]: t1.class_func()

 提示　从输出结果看，Teacher 类已经继承了其父类 Person 的方法 class_func()。

Out [9]: I live in CHINA

In [10]: t1.static_func(1,10)

Out [10]: 11

In [11]: Person.static_func(2,10)

Out [11]: 12

In [12]: t1._deposit

 提示　子类可以继承父类的 protected 属性，如 In[12]中的属性 _deposit。

Out [12]: 100000000000.0

In [13]:
```
t1.__gender #报错信息如下：AttributeError: 'Teacher' object has no attribute '__gender'
```

 提示 子类不能继承父类的private属性，如In[2]中的__gender。

Out [13]:
```
---------------------------------------------------------
AttributeError                            Traceback (most recent call last)
<ipython-input-13-d2490a499a72>in <module>()
----> 1 t1.__gender #报错信息如下：AttributeError: 'Teacher' object has no attribute '__gender'
      2 #【提示】子类不能继承父类的private属性__gender

AttributeError: 'Teacher' object has no attribute '__gender'
```

In [14]:

 提示 可以用以下方法查看类Teacher及其父类Person的docString。

```
Person?
Teacher?
```

In [15]:

 提示 查看类的__name__属性(类名)。注意：Python中的每个类有很多内置的默认属性，其属性名的特点为：以两个下画线开始且以两个下画线结束。比较常用的属性如下。

```python
# __name__：获取类名
# __doc__：获取类的文档字符串，默认为类代码块中第一行的代码块
# __bases__：获取类的所有父类组成的元组
# __dict__：获取类的所有属性和方法组成的列表
# __module__：获取类定义所在的模块名
# __class__：获取实例对应的类

Person.__name__
```

Out [15]:
```
'Person'
```

31.4 私有属性及@property装饰器

In [16]:

> **提示** 与Java和C++不同,Python私有变量的定义语法不用关键字private,而用"两个下画线"来表示私有变量。

```
class Student:
    __name = "Zhang"
```

> **注意** __name为私有变量,但age不是私有变量。

```
    age = 18
    @property
    def get_name(self):
```

> **注意** 此处如果不写 self,则报参数不匹配错误 TypeError: get_name() takes 0 positional arguments but 1 was given。

```
        print(self.__name)
```

> **注意** 此处如果不写self,报错 NameError: name '_Student__name' is not defined。

> **注意** 私有变量的调用:既不能通过类名,也不能通过实例调用。

```
stdnt1 = Student()
```

> **提示** @property装饰器:将方法或函数以属性的形式调用。

```
stdnt1.get_name
```

> **注意** 用Property装饰器修饰的函数的调用不能加"()",必须通过属性方式调用,否则提示错误 TypeError: get_name() takes 0 positional arguments but 1 was given。

Out [16]:   Zhang

31.5 self 和 cls

In [17]:

注意 在定义一个类时,self 代表"实例的引用",如常用于 __init__();cls 代表"类的引用",如常用于__new__()。

```
class Student:
    age = 0
    name = "z"
    def __init__(self):
```

注意 self 只能出现在形式参数中,不能出现在前两行 age 和 name 之前,它是类变量,可以通过类调用。

```
        self.name = "zhang"
```

注意 self.name 是实例变量,与另一个同名类变量 name 不同。

```
        age = 10
```

提示 此处的 age 是 __init__()内的局部变量。

```
s1 = Student()
s2 = Student()
s1.name = "song"
s1.age = 30
Student.age = 20
```

提示 age 为类属性,name 为类属性。

```
Student.name = "li"
```

提示 在 Python 面向对象编程中,类属性(如 name)和实例属性(如 age)在内存中分别占有自己的独立存储空间(互不影响),而实例属性的寻找规则为"先到实例属性中找相应属性,如果在实例属性对应内存中找不对应属性(如 s2.age),则以类属性值代替同名实例属性值"。建议读者采用 Python 通用属性.__dic__跟踪每个类和实例的属性及其属性值,如 s1.__dict__或 Student.__dict__。

```
s1.name, s1.age, s2.name, s2.age
```

Out [17]:　　('song', 30, 'zhang', 20)

31.6 new 与 init 的区别和联系

In [18]:

 思路 通过分析以下代码，可以深入理解 Python 中的 __new__()函数与 __init__()函数的区别以及对象与实例的区别。

```
class Student:
    name = "wang"
    __age = 16

    def __new__(cls, name, age):
        print('new函数被调用')

    def __init__(self, name, age):
        print('init函数被调用')
        self.name = name
        self.age = age
    def sayHi(self):
        print(self.name, self.age)

s1 = Student("zhang", 18)
```

Out [18]:　　new 函数被调用

In [19]:
```
print(s1)
```

 提示 因为 __new__()中没有 return 语句，所以导致 s1 的值为 NoneType。

 思路 Python 中的 __new__()与 __init__()的联系：__new__()中的 return 语句执行后才执行 __init__()。
执行函数 __new__()后生成的是"对象"；执行函数 __init__()后生成的是"实例"。

Out [19]:　　None

In [20]:
```
s1.sayHi()
```

 提示 | 报错：AttributeError: 'NoneType' object has no attribute 'sayHi'

Out [20]:
```
---------------------------------------------------------
AttributeError                      Traceback (most recent call last)
<ipython-input-20-73bde43db512> in <module>()
----> 1 s1.sayHi()
      2 #【提示】报错, AttributeError: 'NoneType' object has no attribute 'sayHi'

AttributeError: 'NoneType' object has no attribute 'sayHi'
```

In [21]:

 提示 | 报错原因分析：问题出现在__new__()函数中没有 return 语句。修改建议：增加 return object.__new__(cls)。

```python
class Student:
    name = "wang"
    __age = 16
    
    def __new__(cls,*args, **kwargs):
        print('new函数被调用')
        return object.__new__(cls)
    
    def __init__(self, name, age):
        print('init函数被调用')
        self.name = name
        self.age = age
    
    def sayHi(self):
        print(self.name, self.age)

s1 = Student("zhang", 18)
s1.sayHi()
```

 思路 请根据如下输出结果理解 Python 中的 __new__()函数与 __init__()函数的区别和联系。

Out [21]:　new 函数被调用
init 函数被调用
zhang 18

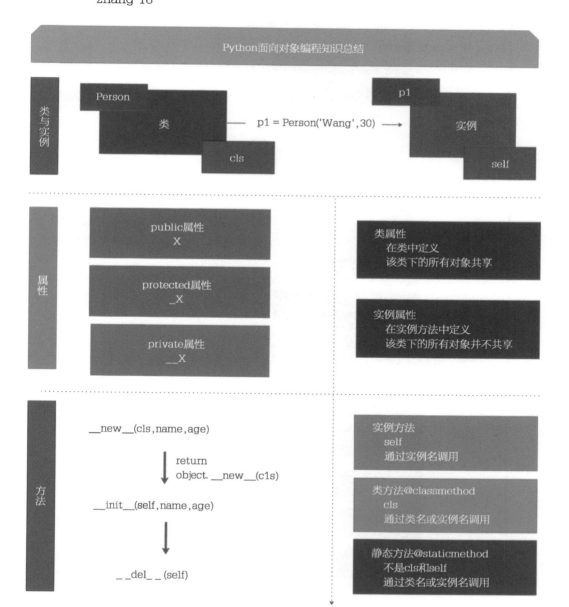

32 魔术命令

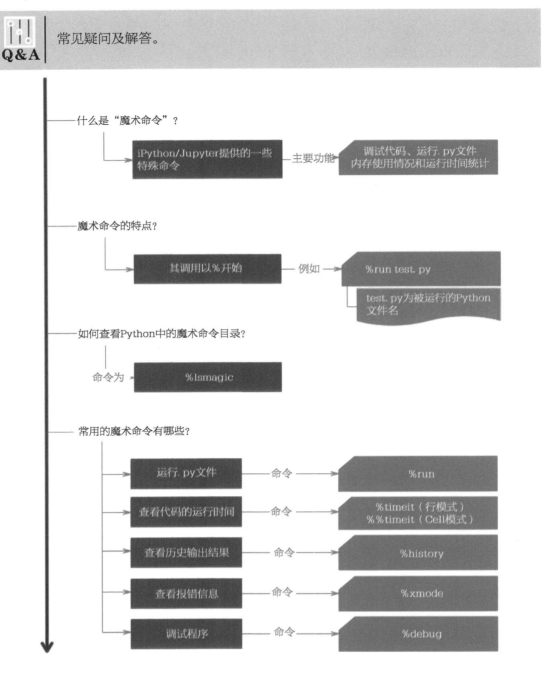

In [1]:

注意 | 魔术命令并非 Python 语言自己的语法，而是 Python 编辑器尤其是基于 iPython 的编辑器（如 iPython、Jupyter Notebook、Jupyter Lab 等）提供的特殊命令，其代码特征为"以%开始"。

32.1 运行.py 文件：%run

In [2]:

提示 | %run 的功能：运行一个用 Python 写的.py 文件（".py"为扩展名的文件为用 Python 编写的源代码文件）。

注意 | 使用前提：事先将.py 文件放在当前工作目录，参见本书【34 当前工作目录（P225/In[1]）】。读者可以从本书配套资料中下载 testme.py 文件。

提示 | 查看当前工作目录的方法——用模块 os 中的函数 getcwd()。

```
import os
os.getcwd()
```

Out [2]:   'C:\\Users\\soloman\\clm'

In [3]:

提示 | 运行.py 文件的方法，以文件"testme.py"为例。

```
%run testme.py
```

Out [3]:   this is testMe.py

In [4]:

注意 | 如果.py 文件并不在"当前工作目录"中，那么 Jupyter Notebook/Lab 将报错。

```
os.chdir("C:/")
```

提示 | os.chdir()的功能为改变当前工作目录，读者应将 teseme.py 文件放在当前工作目录中。

```
%run testme.py
```

32 魔术命令

 报错:
ERROR:root:File ``testme.py`` not found.

ERROR:root:File ``testme.py`` not found.

32.2 统计运行时间：%timeit 与 %%timeit

In [5]:

 （1）查看某行代码（又称为"Line Mode"）的运行时间：
%timeit

 %timeit 的用法：在代码行前加 "%timeit"。

%timeit myList2 = [n**2 for n in range(100)]

 查看帮助信息的方法：
%timeit?

Out [5]: 29.1 μs ± 697 ns per loop (mean ± std. dev. of 7 runs, 10000 loops each)

In [6]: %timeit x = 1

Out [6]: 12.4 ns ± 0.191 ns per loop (mean ± std. dev. of 7 runs, 100000000 loops each)

In [7]:

 （2）查看多行代码（又称为"Cell Mode"）的运行时间：
%%timeit

In [8]: %%timeit x = 1

 与"Line Mode"不同的是，"Cell Mode"中采用双百分号"%%"。除了%%timeit，还可以用更简单方法查看运行时间，即 %%time

```
x = x + 2
x = x * 2
```

Out [8]:   3.23 μs ± 76.2 ns per loop (mean ± std. dev. of 7 runs, 100000 loops each)

In [9]:

 %%timeit 之前不能有任何代码，注释类的代码都不行，否则报错：SyntaxError: invalid syntax。

```
%%timeit x = 1
x = x + 2
x = x * 2
```

Out [9]:   File "<ipython-input-8-10c461c2a871>", line 2
              %%timeit x = 1
              ^
           SyntaxError: invalid syntax

32.3 查看历史 In 和 Out 变量：%history

In [10]:

 %history 的功能：查看历史输出结果，即 Out[]变量。

%history -n 2-3

 显示 In[2]~ In[3]。
%run 的功能：运行.py 文件（用 Python 编写的源代码文件）。
使用前提：已将.py 文件放在当前工作目录中，参见本书【34 当前工作目录（P225/In[1]）】。
查看当前工作目录的方法：用模块 os 中的 getcwd()。

```
import os
os.getcwd()
```

32 魔术命令

 运行 .py 文件的方法，以文件 "testme.py" 为例。

%run testme.py

In [11]: %history –n –O 2

 此处，"–O" 代表的是 Out[]，即 "–O 2" 表示显示 Out[2]，详见帮助信息–%history?。

Out [11]: 2:
'C:\\Users\\soloman\\clm'

 与 Line Mode 不同的是，Cell Mode 采用双百分号 "%%"。

x = x + 2
x = x * 2

32.4 更改异常信息的显示模式：%xmode

In [12]: %xmode 功能：更改错误/异常信息的"显示模式"，参见本书【30 程序调试方法（P194/In[3]）】。

%xmode Plain

 在 Plain 的位置还可以写 Context 或 Verbose，区别在于异常信息的"显示模式"不同。

x1 = 2
x2 = X1

 报错信息为 NameError: name 'X1' is not defined。原因：X1 是未定义变量，之前定义的是 x1。

Out [12]: Exception reporting mode: Plain

Traceback (most recent call last):

  File "<ipython-input-11-58b5577b33b9>", line 7, in <module>
    x2=X1

> 报错：NameError: name 'X1' is not defined。原因：大写 X1 是未定义变量，之前定义的是小写 x1。

NameError: name 'X1' is not defined

In [13]:

> %xmode 功能：更改错误与异常信息的显示模式，参见本书【30 程序调试方法（P194/In[3]）】。

```
%xmode Verbose
```

> 在 Plain 的位置还可以写 Context（默认）、Verbose，区别在于异常信息的显示模式不同。

```
x1 = 2
x2 = X1
```

> 异常信息的报告模式发生了变化，系统提示：
> Exception reporting mode: Verbose

Out [13]: Exception reporting mode: Verbose

NameError   Traceback (most recent call last)
<ipython-input-12-1f8b9c0a8b45>in <module>()
    5 #【提示】在 Plain 的位置还可以写 Context（默认）、Verbose，区别在于异常信息的显示模式不同。
    6 x1 = 2
----> 7 x2 = X1
      global x2 = undefined
          global X1 = undefined
    8

NameError: name 'X1' is not defined

32 魔术命令

32.5 调试程序：%debug

In [14]:
> 提示 | %debug 的功能：打开 Python 调试器（Python Debugger, PDG），参见本书【30 程序调试方法（P193/In[2]）】。
>
> %debug
>
> 注意 | 在系统提示的"ipdb>"输入框中填写要观察的变量名。如果需要退出，那么输入 quit 或 q。
>
> x1 = 2
> x2 = X1

Out [14]: > <ipython-input-12-1f8b9c0a8b45>(7)<module>()
    4 get_ipython().run_line_magic('xmode', 'Verbose')
    5 #【提示】在 Plain 的位置还可以写 Context（默认）、Verbose，区别在于异常信息的显示模式不同。
    6 x1 = 2
----> 7 x2 = X1
    8

ipdb> x1
2
ipdb> X
*** NameError: name 'X' is not defined
ipdb> X1
*** NameError: name 'X1' is not defined
ipdb> x1
2
ipdb> quit

NameError   Traceback (most recent call last)
<ipython-input-13-27b71680f589> in <module>()
  5
  6 x1 = 2
----> 7 x2 = X1
   global x2 = undefined
    global X1 = undefined

NameError: name 'X1' is not defined

32.6 程序运行的逐行统计：%prun 与 %lprun

In [15]:
> 提示 | %prun 与 %lprun 的功能：程序运行情况的逐行统计。
>
> ```
> def myfunc1(n):
>   n = n + 1
>   for i in [1,2,3,4,5]:
>     n = n + i
>   return(n)
> ```

In [16]:
> %prun myfunc1(100)
>
> 提示 | %prun 为 iPython 自带命令。

In [17]:
> %lprun myfunc1(100)
>
> 提示 | 与 %prun 不同的是，%lprun 不是 iPython 自带命令。
>
> 注意 | 报错：UsageError: Line magic function '%lprun' not found。
> 原因：%lprun 不是 iPython 自带命令。

Out [17]: UsageError: Line magic function '%lprun' not found.

In [18]:
> 提示 | 纠正方法：在使用 %lprun 前必须通过 PIP 工具或 Conda 工具安装包 line_profiler，如在命令行中输入 conda install line_profiler，参见本书【27 包（P175/In[2]）】。
>
> 注意 | 目前，lprun 只支持另一个独立 Module（模块）中的函数，不支持本文件中的函数。为此，我们在 testme.py 文件中定义了一个函数 myfunc2()。
>
> ```
> %load_ext line_profiler
> import testme
> ```

32 魔术命令

testme.py 文件需要放在 Anaconda 的 site-packages 文件夹下，site-packages 文件夹的路径为：
~Anaconda\Lib\site-packages。

%lprun -f testme.myfunc2 testme.myfunc2(10)

Out [18]: this is testMe.py

In [19]:

本书的 testme.py 文件的内容如下。

```
print("this is testMe.py")
def myfunc2(n):
    n = n + 1
    for i in[1,2,3,4,5]:
        n = n + i
    return(n)
```

```
Filename: C:\Users\soloman\clm\testme.py
Line #    Mem usage    Increment   Line Contents
================================================
    2     67.7 MiB     67.7 MiB    def myfunc2(n):
    3     67.7 MiB      0.0 MiB        n=n+1
    4     67.7 MiB      0.0 MiB        for i in [1,2,3,4,5]:
    5     67.7 MiB      0.0 MiB            n=n+i
    6     67.7 MiB      0.0 MiB        return(n)
```

注意：testme.py 文件的后缀为 "py"，而不是 Jupyter Notebook 的默认后缀 "ipynb"。

32.7 内存使用情况的统计：%memit

In [20]:

%memit 的功能——内存使用情况的统计。

首先，安装包 memory_profiler，在 Anaconda Prompt 中输入：pip install memory_profiler，参见本书【27 包（P175/In[2]）】。

In [21]:

其次，导入包，在 Jupyter Notebook 中使用另一个魔术命令 "%load_ext"。

%load_ext memory_profiler

In [22]:

 提示   最后，调用%memit 函数和%mprun 函数。

```
%memit myfunc2(100)
```

```
(base) C:\Users\soloman>pip install memory_profiler
Collecting memory_profiler
  Downloading memory_profiler-0.52.0.tar.gz
Requirement already satisfied: psutil in c:\anaconda\lib
Building wheels for collected packages: memory-profiler
  Running setup.py bdist_wheel for memory-profiler ... do
  Stored in directory: C:\Users\soloman\AppData\Local\pip
8f74b67dac
Successfully built memory-profiler
Installing collected packages: memory-profiler
Successfully installed memory-profiler-0.52.0
```

Out [22]:   peak memory: 67.74 MiB, increment: 0.02 MiB

In [23]:

 注意   目前，mprun只能支持另一个独立模块（module）中的函数，不支持本文件中的函数。为此，我们在testme.py文件中定义了一个函数myfunc2()。

```
import testme
%mprun –f testme.myfunc2 testme.myfunc2(10)
```

```
Filename: C:\Users\soloman\clm\testme.py

Line #    Mem usage    Increment   Line Contents
================================================
     2     67.7 MiB     67.7 MiB   def myfunc2(n):
     3     67.7 MiB      0.0 MiB       n=n+1
     4     67.7 MiB      0.0 MiB       for i in [1,2,3,4,5]:
     5     67.7 MiB      0.0 MiB           n=n+i
     6     67.7 MiB      0.0 MiB       return(n)
```

 扩展   Python 数据科学竞赛/开源实践相关的重要网站

Python 数据科学竞赛/开源实践相关的重要网站，建议读者关注。
❖ Kaggle：https://www.kaggle.com/
❖ Kdnuggets：https://www.kdnuggets.com/
❖ GitHub：https://github.com/
❖ UCI 机器学习数据集：https://archive.ics.uci.edu/ml/datasets.html

33 搜索路径

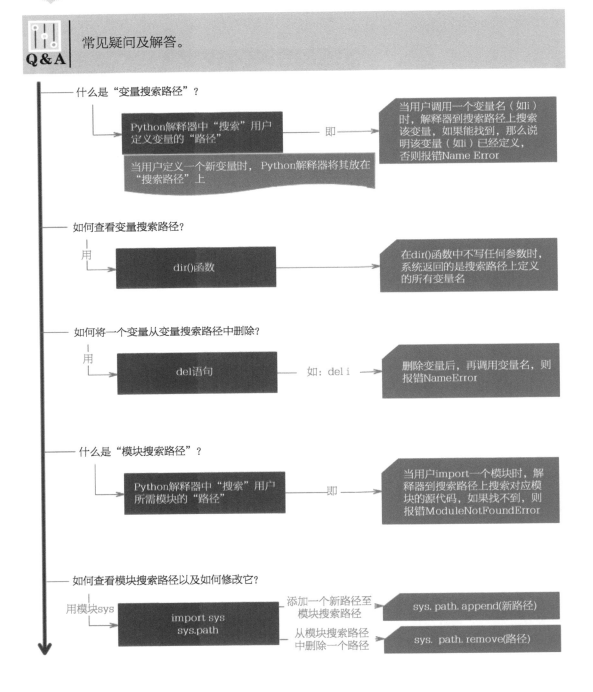

33.1 变量搜索路径

In [1]:
> 提示：查看解释器的"搜索路径"中已存在的所有变量的方法：用内置函数 dir() 或魔术命令 %whos 和 %who。

```
dir()
```

Out [1]: ['In',
'Out',
'_',
'__',
'___',
'__builtin__',
'__builtins__',
'__doc__',
'__loader__',
'__name__',
'__package__',
'__spec__',
'_dh',
'_i',
'_i1',
'_ih',
'_ii',
'_iii',
'_oh',
'exit',
'get_ipython',
'quit']

In [2]:
> 提示：将某个变量放入搜索路径的方法：用赋值语句形式定义一个新变量。例如：

```
vi = 1
```

In [3]:
> 提示：再次显示搜索路径，查看刚定义的新变量"vi"是否已出现在搜索路径之中。

```
dir()
```

33 搜索路径

```
Out [3]:    ['In',
            'Out',
            '_',
            '_1',
            '__',
            '___',
            '__builtin__',
            '__builtins__',
            '__doc__',
            '__loader__',
            '__name__',
            '__package__',
            '__spec__',
            '_dh',
            '_i',
            '_i1',
            '_i2',
            '_i3',
            '_ih',
            '_ii',
            '_iii',
            '_oh',
            'exit',
            'get_ipython',
            'quit',
            'vi']
```

In [4]:

 提示 将某个变量从搜索路径中删除的方法：用 del 语句。

```
del vi
```

 提示 删除变量 vi。

In [5]:

 思路 在 Python 数据分析和数据科学项目中，变量名未定义错误（NameError）出现的根本原因是"搜索路径中找不到它"。

```
vi
```

>
> **提示** 报错：NameError: name 'vi' is not defined。
> 原因分析：在 In[4]中已删除了变量 vi。

Out [5]:
```
------------------------------------------
NameError                                Traceback (most recent call last)
<ipython-input-5-cf2c6cd2c98b> in <module>()
----> 1 vi #报错 NameError: name 'vi' is not defined

NameError: name 'vi' is not defined
```

33.2 模块搜索路径

In [6]:
>
> **思路** 模块搜索路径的查看方法——sys 模块中提供的属性 path。此外，还可以通过在 Anaconda Prompt 中采用输入命令 python -m site 方式查看模块搜索路径。

```python
import sys
sys.path
```

>
> **注意** sys.path 是属性，而不是方法，不能加括号。

Out [6]:   ['',
 'C:\\Anaconda\\python36.zip',
 'C:\\Anaconda\\DLLs',
 'C:\\Anaconda\\lib',
 'C:\\Anaconda',
 'C:\\Anaconda\\lib\\site-packages',
 'C:\\Anaconda\\lib\\site-packages\\win32',
 'C:\\Anaconda\\lib\\site-packages\\win32\\lib',
 'C:\\Anaconda\\lib\\site-packages\\Pythonwin',
 'C:\\Anaconda\\lib\\site-packages\\iPython\\extensions',
 'C:\\Users\\soloman\\.iPython']

In [7]:
>
> **思路** 在 Python 中，增加一个新路径至模块搜索路径的方法为：
> sys.path.append()

33 搜索路径

```
import sys
sys.path.append('H:\\Python\\Anaconda')
```

In [8]:
```
sys.path
```

 提示 | 再次显示模块搜索路径,查看在 In[7]中新增的路径是否已出现在模块搜索路径之中。

Out [8]:
```
[''',
 'C:\\Anaconda\\python36.zip',
 'C:\\Anaconda\\DLLs',
 'C:\\Anaconda\\lib',
 'C:\\Anaconda',
 'C:\\Anaconda\\lib\\site-packages',
 'C:\\Anaconda\\lib\\site-packages\\win32',
 'C:\\Anaconda\\lib\\site-packages\\win32\\lib',
 'C:\\Anaconda\\lib\\site-packages\\Pythonwin',
 'C:\\Anaconda\\lib\\site-packages\\iPython\\extensions',
 'C:\\Users\\soloman\\.iPython',
 'H:\\Python\\Anaconda']
```

In [9]:

 思路 | 从模块搜索路径中删除一个路径:
sys.path.remove()

```
sys.path.remove('H:\\Python\\Anaconda')
```

In [10]:
```
sys.path
```

 提示 | 再次显示模块搜索路径,查看在 In[9]中已删除的路径是否已不再显示在模块搜索路径中。

Out [10]:
```
[''',
 'C:\\Anaconda\\python36.zip',
 'C:\\Anaconda\\DLLs',
 'C:\\Anaconda\\lib',
 'C:\\Anaconda',
 'C:\\Anaconda\\lib\\site-packages',
```

```
'C:\\Anaconda\\lib\\site-packages\\win32',
'C:\\Anaconda\\lib\\site-packages\\win32\\lib',
'C:\\Anaconda\\lib\\site-packages\\Pythonwin',
'C:\\Anaconda\\lib\\site-packages\\iPython\\extensions',
'C:\\Users\\soloman\\.iPython']
```

扩展 | R 语言与 Python 的对照。

	R 语言	Python
设计者	Ross Ihaka 和 Robert Gentleman（统计学家）	Guido Van Rossum（程序员）
设计目的	方便统计处理、数据分析及图形化显示	提升软件开发的效率和源代码的可读性
设计哲学	简单、有效、完善（功能层次上）	优雅、明确、简单（源代码层次上）
发行年	1995	1991
前身	S 语言	ABC 语言、C 语言和 Modula-3 语言
主要维护者	The R-Core Team（R-核心团队） The R Foundation（R 基金会）	Python Software Foundation（Python 软件基金会）
第三方提供的功能	可从 CRAN 下载	可从 PyPi 下载
常用包/库	数据科学工具集：tidyverse 数据处理：dplyr, plyr, data.table, stringr 可视化：ggplot2, ggvis, lattice 机器学习：RWeka, caret	数据处理：pandas 科学计算：SciPy, NumPy 可视化：matplotlib 统计建模：statsmodels 机器学习：sckikit-learn, TensorFlow, Theano

34 当前工作目录

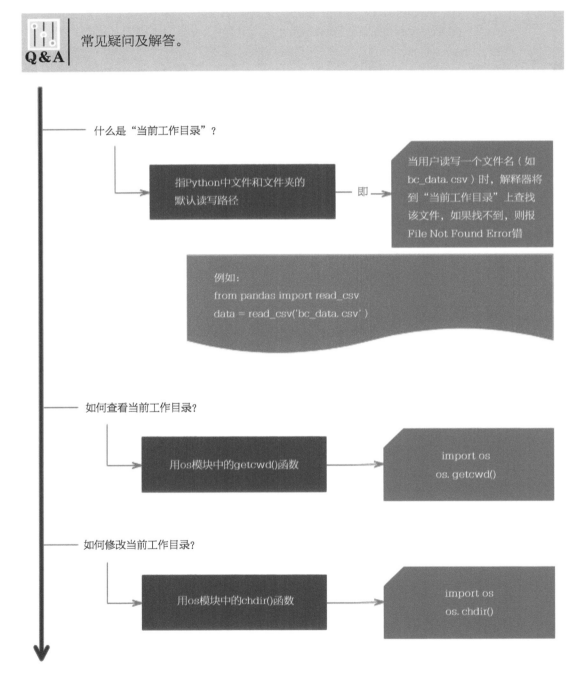

In [1]:
> 提示 "当前工作目录"就是Python（准确地说，Python解释器和编辑器，如Jupyter Notebook）的读写文件和文件夹的默认路径。例如，本书【38 DataFrame】中需要读取外部文件"bc_data.csv"，应事先把该文件放在"当前工作目录"中。

34.1 显示当前工作目录的方法

In [2]:
> 提示 用模块 os 的方法：
> getcwd()

```
import os
print(os.getcwd())
```

Out [2]:   C:\Users\soloman\clm

34.2 更改当前工作目录的方法

In [3]:
> 提示 更改当前工作目录的方法：用模块 os 中的函数 chdir()。在此，先用函数 getcwd()查看当前工作目录。

```
os.getcwd()
```

> 注意 更改当前工作目录的前提是"已创建了即将用作当前工作目录的文件夹"，如 E:\PythonProjects。

```
os.chdir('E:\PythonProjects')
print(os.getcwd())
```

> 提示 更改当前工作目录后，仅在"当前会话"中有效，详见本书【3 如何看懂和执行本书代码】。

Out [3]:   E:\PythonProjects

34.3 读、写当前工作目录的方法

In [4]:

 数据分析师应根据目标数据文件的类型和数据分析工作的需要，采用不同的导入方法，如用内置函数 open()或第三方扩展包 Pandas 的 read_csv()、read_excel()等。

 例如，将当前工作目录下的文件"bc_data.csv"读入本地数据框"data"。

```
from pandas import read_csv
data = read_csv('bc_data.csv')
data.head(5)
```

 read_csv('bc_data.csv')的前提是将目标文件"bc_data.csv"已放在当前工作目录中，如"E:\PythonProjects"。

 读者可以从本书提供的配套资源中找到数据文件"bc_data.csv"。

Out [4]:

	id	diagnosis	radius_mean	texture_mean	perimeter_mean	area_mean	smooth
0	842302	M	17.99	10.38	122.80	1001.0	0.11840
1	842517	M	20.57	17.77	132.90	1326.0	0.08474
2	84300903	M	19.69	21.25	130.00	1203.0	0.10960
3	84348301	M	11.42	20.38	77.58	386.1	0.14250
4	84358402	M	20.29	14.34	135.10	1297.0	0.10030

5 rows × 32 columns

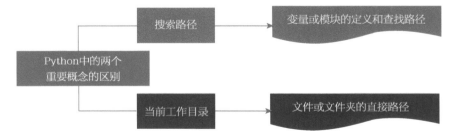

第四篇
数据加工

随机数
数组
Series
DataFrame
日期与时间
可视化
Web 爬取

35 随机数

 常见疑问及解答。

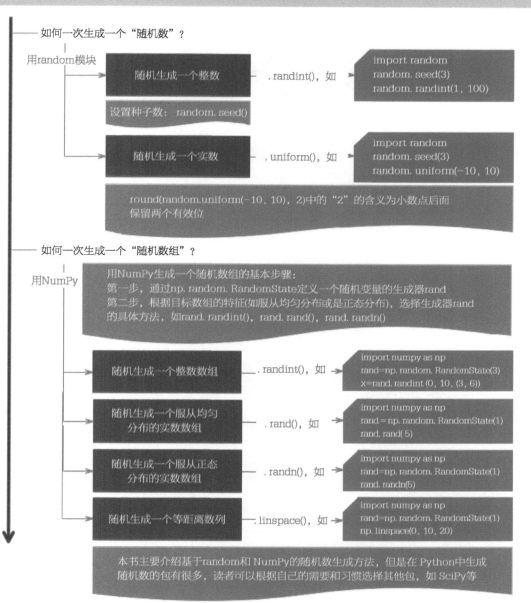

35 随机数

35.1 一次生成一个数

In [1]:

 一次生成一个随机数的方法：用 random 模块。

```
import random
random.seed(3)
```

 random.seed()的功能为生成随机数的"种子数"。

 Python中实现同一种功能的包/模块会有很多，不同的包/模块在用户体验、实现技术、优化程度和主要应用场景有所不同。例如，random并不是Python中生成随机数的唯一的包，还有NumPy、SciPy等。

In [2]:

 生成 [1, 100]之间一个的随机整数。

```
import random
random.seed(3)
random.randint(1, 100)
```

 random.seed(3)的含义为"设置随机数的种子数"，如果不设置，那么系统每次生成的随机数都不一样。

Out [2]:   31

In [3]:

 生成 [−10, 10]之间的一个随机浮点数（实数）。

```
import random
random.seed(3)
random.uniform(−10, 10)
```

Out [3]:   −5.240707458162173

In [4]:
> 建议查看 random.uniform 的说明文档，学习更多内容。

random.uniform?

In [5]:
> random.seed(3)
> round(random.uniform(-10, 10),2)
>
> 在 "round(random.uniform(-10, 10),2)" 中，"2" 的含义为小数点后面保留两个有效位。

Out [5]: −5.24

35.2 一次生成一个随机数组

In [6]:
> 用包 NumPy（参见本书【36 数组（P238/In[1]）】）生成一个随机数组的基本步骤如下。
>
> #第一步，通过np.random.RandomState定义一个随机变量的生成器rand
> #第二步，根据目标数组的特征（如服从均匀分布还是正态分布），选择生成器 rand 提供的具体方法，如 rand.randint()、rand.rand()、rand.randn()

In [7]:
> （1）生成整数，如生成一个 3×6 的矩阵，矩阵的每个元素为 [0,10] 之间的整数。
>
> import numpy as np
> rand = np.random.RandomState(32)
>
> 此处，np.random.RandomState()的参数 32 为随机数种子数，(3,6)为目标数组的形状。
>
> x = rand.randint(0,10,(3,6))
> x

Out [7]: array([[7, 5, 6, 8, 3, 7],
       [9, 3, 5, 9, 4, 1],
       [3, 1, 2, 3, 8, 2]])

35 随机数

In [8]:

 提示 （2）生成服从均匀分布的浮点数（实数）数组，如生成一个含有 5 个元素的数组。

```
import numpy as np
rand = np.random.RandomState(1)
x = rand.rand(5) *10
```

 提示 rand.rand()的返回值的取值范围为[0,1]，在此"*10"的目的是"调整所生成随机数的取值范围"。

x

Out [8]: array([4.17022005e+00, 7.20324493e+00, 1.14374817e−03, 3.02332573e+00, 1.46755891e+00])

In [9]:

 提示 （3）生成服从正态分布的浮点数（实数）数组，如生成一个含有 5 个元素且服从正态分布的数组。

```
import numpy as np
rand = np.random.RandomState(1)
y = rand.randn(5) +5
```

 提示 请读者思考 In[8]和 In[9]中分别对 rand.rand(5) 进行"*10"和"+5"的区别及应用场景是什么？

y

Out [9]: array([6.62434536, 4.38824359, 4.47182825, 3.92703138, 5.86540763])

In [10]:

 提示 （4）生成等距数列，如产生一个含有 20 个元素的等距数列，其中每个元素的取值范围为[0,10]。

x = np.linspace(0,10,20)

 注意 此处用的是 NumPy 包中的 np.linspace()。生成等比数列的函数为 np.logspace(0,10,20)，更多函数及其用法请参见 NumPy 的官方文档。

x

```
Out [10]:   array([ 0.        ,  0.52631579,  1.05263158,  1.57894737,  2.10526316,
            2.63157895,  3.15789474,  3.68421053,  4.21052632,  4.73684211,
            5.26315789,  5.78947368,  6.31578947,  6.84210526,  7.36842105,
            7.89473684,  8.42105263,  8.94736842,  9.47368421, 10.        ])
In [11]:
```

> **技巧** 建议读者查阅 np.linspace 的说明文档，了解更多内容。

```
np.linspace?
```

> **扩展** 女性不适合从事数据科学和大数据分析职业？

非也。2018 年，Onalytica 公布了一份名为《2018 Data Science Influencer Report （2018 数据科学影响力报告）》的研究报告，给出了数据科学领域的 100 个顶级影响力人物、知名品牌和出版物。其中的亮点是：除了排名第一的 IBM，建议大家关注排名第二的 Girls Who Code（直译：写代码的女孩们）。该品牌是 Reshma 发起的一个非常有想法的项目。用 Reshma 的话讲，"Girls Who Code was founded with a single mission: to close the gender gap in technology."，WOMEN IN DATA SCIENCE 首页如图 35-1 所示。

据本书作者观察，数据科学是性别歧视最小的领域之一，很多女性在引领这一新领域，如 Mason、Cortes、Koller、Monica 都是世界顶级的数据科学家。

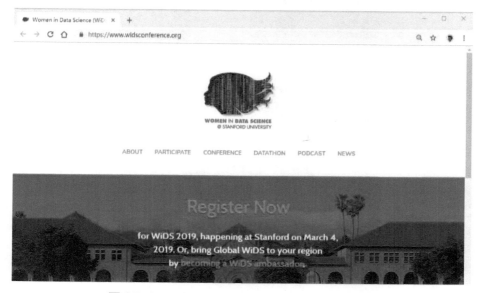

图 35-1　WOMEN IN DATA SCIENCE 首页

36 数组

 常见疑问及解答。

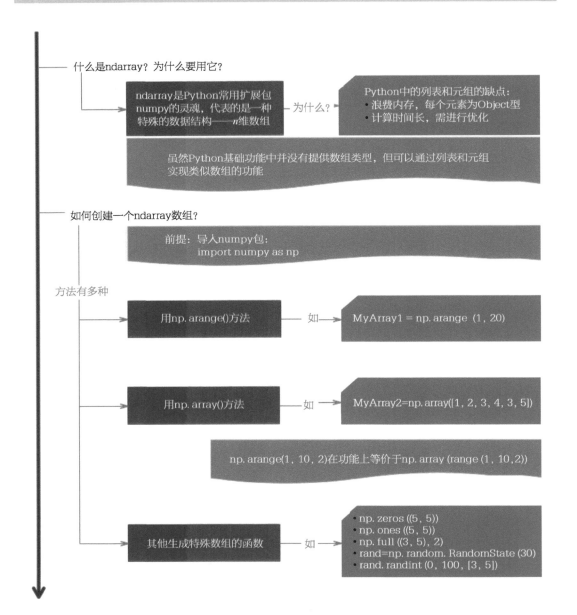

- ndarray数组有什么特殊性？

  - shape参数 —代表的是→ 数组的形状，即几行几列
    - 数组的重构，即改变数组的形状，用reshape()方法，如MyArray5.reshape(4, 5)
    - 注意reshape()与resize()的区别

  - dtype参数 —代表的是→ 数组元素的数据类型

  - ufunc()函数 —代表的是→ 以列为单位计算的函数，目的是支持向量计算，不用写循环语句
    - ufunc()函数是ndarray数组中的多数方法的共性特点，即支持向量计算

- 如何访问ndarray数组中的元素？

  - ndarray的切片和读取，与列表类似，参见本书【14 列表】

  - 若需要访问的index有规则，则采用类似列表的方法 —如→ myArray[1:5:2]

  - 若需要访问的元素index没有规则，则采用FancyIndexing方法 —如→ myArray[[1, 3, 6]]
    - FancyIndexing：将需要访问的元素下标以列表形式提供

36 数组

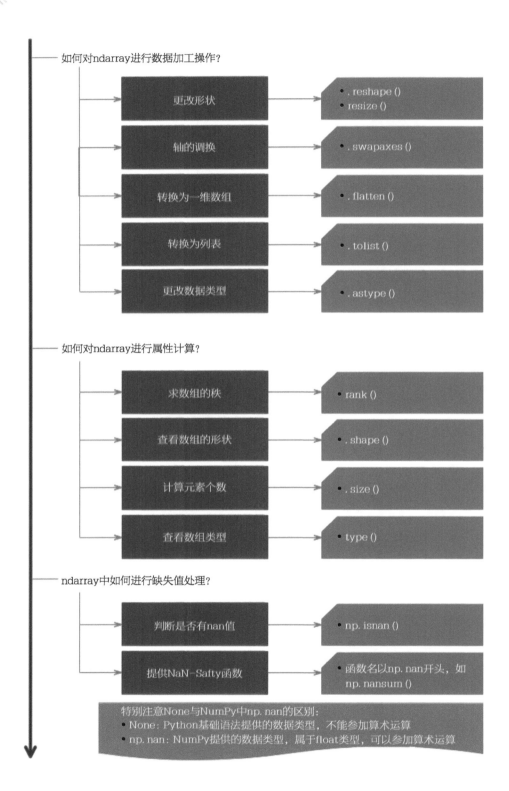

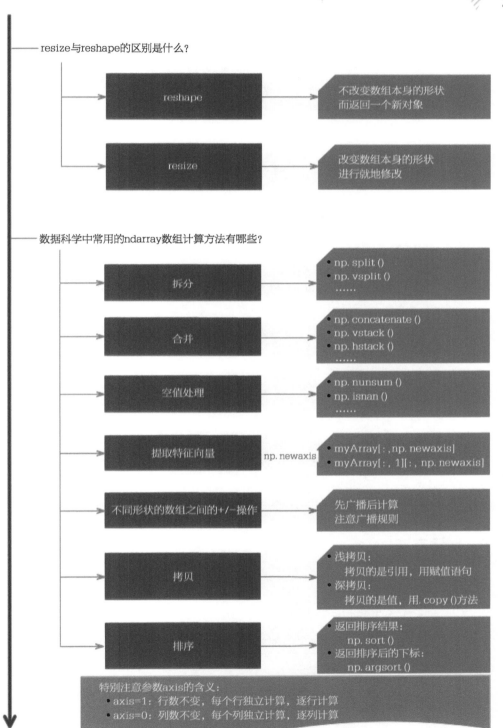

36 数组

In [1]:

思路 | 关于 Python 中的"数组",读者需要了解以下背景知识。

（1）Python 的基础语法中并没有提供数组类型，数组的功能可以用列表和元组来实现。

（2）用列表或元组来代替 C 和 Java 中的数组的缺点——时间和空间代价都很大。原因：在 Python 中，列表和元组中的每个元素都是按"对象"来处理的，每个成员都需要存储引用和对象值。

（3）Python 中出现了"以优化列表和元组，进而实现数组功能"的第三方扩展包，如 NumPy 等。注：另有一个扩展模块叫 array，但一般不用它。

（4）最常用的是 Python 扩展包 NumPy 中的数据结构 ndarray。相对于列表和元组，ndarray 有三个优点：更节省内存、更节省运行时间、更方便使用。原因：NumPy 是用 C 语言实现的，并进行了一定的优化处理。

（5）ndarray 的本质是 n 维数组。
特殊性体现在"支持通过参数 dtype 设置数组元素的类型"。

In [2]:

提示 | 在调用 ndarray 之前需要导入 NumPy 模块。关于模块的导入方法参见本书【26 模块（P170/In[3]）】。

```
import numpy as np
```

36.1 创建方法

In [3]:

思路 | ndarray 对象的创建方法有多种。

In [4]:

提示 | 第一类方法：用 np.arange()。

> **注意** ndarray 中有很多自带函数，可以用于创建不同类型的数组，如 arange()、array()、ones_like()、zeros()、ones()等。

MyArray1 = np.arange(1,20)

> **提示** np.arange(1,20)的功能与range(1,20)类似，返回一个由大于等于1（包含1）且小于20（不包含20）的自然数组成的有序数组。

MyArray1

> **注意** range()与 arange()的区别。

#range 是 Python 内置函数，返回值为 range 对象，见本书【11 for 语句】。
#arange 是扩展模块 numpy 中的函数，返回值为 NumPy 的 ndarray。

Out [4]: array([1, 2, 3, 4, 5, 6, 7, 8, 9, 10, 11, 12, 13, 14, 15, 16, 17, 18, 19])

In [5]: range(1,10,2)

> **提示** 从输出结果看，range()的返回值为一个迭代器，详见本书【25 迭代器与生成器（P165）】。

Out [5]: range(1, 10, 2)

In [6]: list(range(1,10,2))

> **提示** 通过强制类型转换即可显示 range()返回的迭代器中的内容。

Out [6]: [1, 3, 5, 7, 9]

In [7]: np.arange(1,10,2)

> **提示** 虽然函数 range()与 np.arange()的功能相同，但后者是 NumPy 的方法，运行速度更快，占用内存更小，用起来更方便。

> **思路** 在数据分析与数据科学项目中，同一个功能的实现方法有很多种（如用 Python 基本语法代码和调用第三方扩展包）。但是，不同方法的时间复杂度、空间复杂度和灵活性不一样。通常，第三方包会对 Python 基础语法进行优化处理。

36 数组

Out [7]: array([1, 3, 5, 7, 9])

In [8]:

 第二类方法：用 np.array()。

MyArray2 = np.array([1,2,3,4,3,5])

 array()是模块 NumPy 中的函数。

MyArray2

Out [8]: array([1, 2, 3, 4, 3, 5])

In [9]:

 np.array(range(1,10,2))等价于 np.arange(1,10,2)。

np.array(range(1,10,2))

 range()是 Python 内置函数，arange()是扩展包 numpy 中的函数。

Out [9]: array([1, 3, 5, 7, 9])

In [10]:

 第三类方法：用 np.zeros()、np.ones()等函数。

MyArray3 = np.zeros((5,5))

 参数(5,5)代表的是目标数组的形状（shape），即 5 行 5 列的数组。

MyArray3

Out [10]: array([[0., 0., 0., 0., 0.],
       [0., 0., 0., 0., 0.],
       [0., 0., 0., 0., 0.],
       [0., 0., 0., 0., 0.],
       [0., 0., 0., 0., 0.]])

 np.arange()的返回值为 array。

In [11]:
```
MyArray4 = np.ones((5,5))
MyArray4
```

Out [11]: array([[1., 1., 1., 1., 1.],
　　　　　　 [1., 1., 1., 1., 1.],
　　　　　　 [1., 1., 1., 1., 1.],
　　　　　　 [1., 1., 1., 1., 1.],
　　　　　　 [1., 1., 1., 1., 1.]])

In [12]:
 提示　第四类方法：用 np.full() 创建相同元素的数组。

```
np.full((3,5),2)
```

Out [12]: array([[2, 2, 2, 2, 2],
　　　　　　 [2, 2, 2, 2, 2],
　　　　　　 [2, 2, 2, 2, 2]])

In [13]:
 提示　第五类方法：用 np.random() 生成随机数组，参见本书【35.2 一次生成一个随机数组（P231/In[6]）】。以下代码中，0 和 100 代表的是随机数的取值范围，"3,5" 代表的是目标数组的形状，即 3 行 5 列。

```
rand = np.random.RandomState(30)
MyArray5 = rand.randint(0,100,[3,5])
MyArray5
```

Out [13]: array([[37, 37, 45, 45, 12],
　　　　　　 [23, 2, 53, 17, 46],
　　　　　　 [3, 41, 7, 65, 49]])

36.2 主要特征

In [14]:
 思路　ndarray 的两个重要特征如下。

```
#（1）shape：多维数组的形状
#其取值为一个元组或列表
#例如，shape = (2,15)代表的是一个2行15列的数组
```

36 数组

```
#（2）dtype：多维数组中的元素的数据类型
#其取值为np.int等numpy模块提供的数据类型
#例如：dtype = np.int，代表的是数组元素为numpy模块中的int型
```

 ndarray的数据类型比Python自带的类型多。

In [15]:
```
import numpy as np
MyArray4 = np.zeros(shape = (2,15),dtype = np.int)
MyArray4
```

 注意事项：

```
#第一，"shape = "字样可以省略
#第二，np.int不加双引号
```

Out [15]: array([[0, 0, 0, 0, 0, 0, 0, 0, 0, 0, 0, 0, 0, 0, 0],
　　　　　　　　[0, 0, 0, 0, 0, 0, 0, 0, 0, 0, 0, 0, 0, 0, 0]])

In [16]: shape参数代表的是数组的形状，取值可以为元组，如(3,5)。

np.ones((3,5),dtype = float)

Out [16]: array([[1., 1., 1., 1., 1.],
　　　　　　　　[1., 1., 1., 1., 1.],
　　　　　　　　[1., 1., 1., 1., 1.]])

In [17]: shape参数的取值也可以为列表，如[3,5]。

np.ones([3,5],dtype = float)

Out [17]: array([[1., 1., 1., 1., 1.],
　　　　　　　　[1., 1., 1., 1., 1.],
　　　　　　　　[1., 1., 1., 1., 1.]])

36.3 切片/读取

In [18]: 思路 | ndarray 的切片/读取操作与列表非常相似，参见本书【14 列表（P76）】。

In [19]: 提示 | 先创建试验数据集 myArray。

```
import numpy as np
myArray = np.array(range(1,10))
myArray
```

Out [19]: array([1, 2, 3, 4, 5, 6, 7, 8, 9])

In [20]:
```
myArray = np.arange(1,10)
```
 提示 | 相当于 myArray = np.array(range(1,10))。

```
myArray
```

Out [20]: array([1, 2, 3, 4, 5, 6, 7, 8, 9])

In [21]: 注意 | （1）第一个元素的下标为 0。

```
myArray[0]
```

Out [21]: 1

In [22]: 注意 | （2）Python 中有负下标，详见本书【14 列表（P76）】。

```
myArray[-1]
```

Out [22]: 9

 注意 | （3）Python 下标的几种写法，详见本书【14 列表（P76）】。

36 数组

In [23]:

```
import numpy as np
myArray = np.array(range(0,10))

print("myArray = ",myArray)
```

 查看数组 myArray 的当前值。

```
print("myArray[1:9:2] = ", myArray[1:9:2])
```

 1、9、2 分别为切片操作的开始位置（start）、结束位置（stop-1）和步长（step）。

```
print("myArray[:9:2] = ", myArray[:9:2])
```

 可以省略 start。

```
print("myArray[::2] = ", myArray[::2])
```

 可以省略 start 和 stop。

```
print("myArray[::] = ", myArray[::])
```

 start、stop 和 step 都可以省略。

```
print("myArray[:8:] = ", myArray[:8:])
```

 可以省略 start 和 stop。

```
print("myArray[:8] = ", myArray[0:8])
```

 可以省略 step。

```
print("myArray[4::] = ", myArray[4::])
```

 可以省略 stop 和 step。

```
print("myArray[9:1:-2] = ", myArray[9:1:-2])
```

```python
print("myArray[::-2] = ", myArray[::-2])
```

 提示 | step 值可以为负数。

```python
print("myArray[[2,5,6]] = ", myArray[[2,5,6]])
```

 提示 | Fancy Indexing 是一个非常灵活的切片技术，其含义为"支持非迭代方式，即不规则读取（切片）元素"，其存在标志为 [] 的嵌套，即 [] 中出现另一个 []。例如，myArray[[2,5,6]] 的含义为读取下标为 2、5 和 6 的三个元素。

```python
print("myArray[myArray>5] = ", myArray[myArray>5])
```

 提示 | 下标中还可能出现含有数组名本身的表达式，含义为"过滤条件"。

 提示 | 除了上述情况，Python 下标中还可以出现 np.newaxis 字样，详见 In[33]。

Out [23]:
```
myArray= [0 1 2 3 4 5 6 7 8 9]
myArray[1:9:2] = [1 3 5 7]
myArray[:9:2] = [0 2 4 6 8]
myArray[::2] = [0 2 4 6 8]
myArray[::] = [0 1 2 3 4 5 6 7 8 9]
myArray[:8:] = [0 1 2 3 4 5 6 7]
myArray[:8] = [0 1 2 3 4 5 6 7]
myArray[4::] = [4 5 6 7 8 9]
myArray[9:1:-2] = [9 7 5 3]
myArray[::-2] = [9 7 5 3 1]
myArray[[2,5,6]] = [2 5 6]
myArray[myArray>5] = [6 7 8 9]
```

In [24]:
```python
myArray[0:2]
```

 注意 | 切片操作中，start 是包含的（如本代码中的"0"），但不包含 stop（如本代码中的"2"），规则为"左包含、右不包含"。

Out [24]: array([0, 1])

36 数组

In [25]: myArray[1:5:2]

 此处为两个冒号,start、stop、step 中的冒号都可以省略。

Out [25]: array([1, 3])

In [26]:
 step 值可以为正数,含义为:从第一个元素开始往后遍历。

myArray[::2]

Out [26]: array([0, 2, 4, 6, 8])

In [27]:
 step 值也可以为负数,含义为:从最后一个元素开始往前遍历。

myArray[::-2]

Out [27]: array([9, 7, 5, 3, 1])

In [28]:
 切片操作时,数组本身不会发生改变。

myArray

Out [28]: array([0, 1, 2, 3, 4, 5, 6, 7, 8, 9])

In [29]:
 (4)数组的非连续元素的读取方法:用"切片",参见本书【14 列表(P76)】。

myArray = np.array(range(1,11))
myArray

 关于 range()函数参见本书【14 列表(P76)】。

Out [29]: array([1, 2, 3, 4, 5, 6, 7, 8, 9, 10])

In [30]:

注意 | 初学者容易忽略的问题：当下标不规则时，不用 Fancy Indexing 则出错，具体参见本书【36 数组】中对 Fancy Indexing 的介绍（P245）。例如：

myArray[1,3,6]

注意 | 报错信息：IndexError: too many indices for array（索引的维度过多的错误信息）。

提示 | 纠错方法：将"[1,3,6]"改为"[[1,3,6]]"，即采用切片。

Out [30]: --

```
IndexError                                Traceback (most recent call last)
<ipython-input-30-13b1cd8a6af6> in <module>()
      1
----> 2 myArray[1,3,6]
      3

IndexError: too many indices for array
```

In [31]:

提示 | 纠正上一条错误的方法——用"切片"的方法，参见本书【14 列表（P76）】。

myArray[[1,3,6]]

技巧 | 采用 Fancy Indexing 方法切片的特点是下标中嵌套出现方括号，如[[ ]]。

Out [31]: array([2, 4, 7])

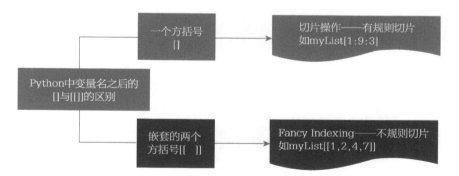

36 数组

In [32]:

 思路 在数据分析和数据科学项目中，通常需要生成一个特殊矩阵，即"特征矩阵"。

myArray

 提示 从输出结果看，myArray 的当前值为一行记录，不符合"特征矩阵的要求"，需要对其进行规整化处理。

Out [32]: array([1, 2, 3, 4, 5, 6, 7, 8, 9, 10])

In [33]:

 思路 规整化处理特征矩阵的方法。

myArray[:,np.newaxis]

 提示 np.newaxis 是指定义一个新维度（列），功能和用法与 None 一样。NumPy 中 ndarray 的下表/索引的每个维度上（即一个逗号的前后或两个逗号之间），只能出现以下内容：整数、冒号（:）、numpy.newaxis 或 None，以及元素为整数或布尔值的数组（[]）。

 注意 此处的冒号（:）不能省略。

Out [33]: array([[1],
　　　　　　　　[2],
　　　　　　　　[3],
　　　　　　　　[4],
　　　　　　　　[5],
　　　　　　　　[6],
　　　　　　　　[7],
　　　　　　　　[8],
　　　　　　　　[9],
　　　　　　　　[10]])

In [34]:

 提示 查看形状用属性 shape。

myArray[:,np.newaxis].shape

Out [34]: (10, 1)

In [35]:
 提示 | 更改形状用 NumPy 中的方法 reshape()。

myArray2 = np.arange(1,21).reshape([5,4])
myArray2

Out [35]: array([[1, 2, 3, 4],
       [5, 6, 7, 8],
       [9, 10, 11, 12],
       [13, 14, 15, 16],
       [17, 18, 19, 20]])

In [36]:
 提示 | 多维数组的切片读取方法示例如下。

myArray2[[2,4],3]

Out [36]: array([12, 20])

In [37]:
 提示 | 另一种更好的写法如下。

x = [2,4]
myArray2[x,3]

Out [37]: array([12, 20])

36.4 浅拷贝和深拷贝

In [38]:
 提示 | 浅拷贝——复制过来的是"引用",即"复制对象和被复制对象共享一个存储空间,二者并不是相互独立的"。

import numpy as np
myArray1 = np.array(range(0,10))
myArray2 = myArray1
myArray2[1] = 100   #此处修改了 myArray2 的取值。
myArray1

36 数组

注意 | 此处，myArray 的取值已发生改变，原因分析："myArray2= myArray1"属于浅拷贝，myArray1 和 myArray2 共享一个存储空间。

Out [38]: array([0, 100, 2, 3, 4, 5, 6, 7, 8, 9])

In [39]:

提示 | 深拷贝——复制过来的是"值"，即"复制对象和被复制对象占用两个不同空间，二者是相互独立的"。

注意 | ndarray 的深拷贝方法：copy()

```
import numpy as np
myArray1 = np.array(range(0,10))
myArray2 = myArray1.copy()
myArray2[1] = 200        #此处修改了 myArray2 的取值。
myArray1
```

提示 | 此处的myArray1没有发生改变，原因分析："myArray2= myArray1.copy()"属于深拷贝，myArray1和myArray2是相互独立的。

Out [39]: array([0, 1, 2, 3, 4, 5, 6, 7, 8, 9])

36.5 形状与重构

In [40]:

提示 | 重构（reshape）的含义为"返回一个符合新形状要求的数组"。

```
import numpy as np
MyArray5 = np.arange(1,21)
MyArray5
```

Out [40]: array([1, 2, 3, 4, 5, 6, 7, 8, 9, 10, 11, 12, 13, 14, 15, 16, 17, 18, 19, 20])

In [41]:
> 查看形状。

MyArray5.shape

Out [41]: (20,)

In [42]:
> (1)用 reshape()返回另一个新的数组。

MyArray6 = MyArray5.reshape(4,5)
MyArray6

> reshape()不会改变数组本身。

Out [42]: array([[1, 2, 3, 4, 5],
       [6, 7, 8, 9, 10],
       [11, 12, 13, 14, 15],
       [16, 17, 18, 19, 20]])

In [43]: MyArray5.shape

Out [43]: (20,)

In [44]: MyArray5

Out [44]: array([1, 2, 3, 4, 5, 6, 7, 8, 9, 10, 11, 12, 13, 14, 15, 16, 17,
       18, 19, 20])

In [45]: MyArray5.reshape(5,4)

> 返回另一个新的 5×4 的二维数组。

Out [45]: array([[1, 2, 3, 4],
       [5, 6, 7, 8],
       [9, 10, 11, 12],
       [13, 14, 15, 16],
       [17, 18, 19, 20]])

In [46]: MyArray5.reshape(5,5)

36 数组

 报错：ValueError: cannot reshape array of size 20 into shape (5,5)。原因分析：reshape 的前提是"可以 reshape"。

Out [46]:
```
---------------------------------------------------
ValueError                    Traceback (most recent call last)
<ipython-input-46-8920a583f59a>in <module>()
----> 1 MyArray5.reshape(5,5)
      2
ValueError: cannot reshape array of size 20 into shape (5,5)
```

In [47]:
```
MyArray5
```

Out [47]: array([1, 2, 3, 4, 5, 6, 7, 8, 9, 10, 11, 12, 13, 14, 15, 16, 17, 18, 19, 20])

In [48]: （2）用 resize() 方法更改数组本身的形状，即"就地修改"。

```
MyArray5.resize(4,5)
MyArray5
```

 resize 与 reshape 的区别：前者修改数组本身（即"就地修改"），后者不修改数组本身（即返回另一个新数组）。

Out [48]: array([[1, 2, 3, 4, 5],
　　　　　　　　　[6, 7, 8, 9, 10],
　　　　　　　　　[11, 12, 13, 14, 15],
　　　　　　　　　[16, 17, 18, 19, 20]])

In [49]: （3）用 swapaxes() 方法进行轴调换，如实现矩阵转置操作。

```
MyArray5.swapaxes(0,1)
```

Out [49]: array([[1, 6, 11, 16],
　　　　　　　　　[2, 7, 12, 17],
　　　　　　　　　[3, 8, 13, 18],
　　　　　　　　　[4, 9, 14, 19],
　　　　　　　　　[5, 10, 15, 20]])

In [50]:

 swapaxes(0,1)不改变数组本身。

MyArray5

 在数据分析和数据科学项目中,需要特别注意对某个数据对象的计算过程是否更改数据本身,还是返回一个新值。

Out [50]: array([[1, 2, 3, 4, 5],
      [6, 7, 8, 9, 10],
      [11, 12, 13, 14, 15],
      [16, 17, 18, 19, 20]])

In [51]:
```
MyArray5 = MyArray5.swapaxes(0,1)
MyArray5
```

Out [51]: array([[1, 6, 11, 16],
      [2, 7, 12, 17],
      [3, 8, 13, 18],
      [4, 9, 14, 19],
      [5, 10, 15, 20]])

In [52]:

 (4)用 flatten()方法将多维数组转换成一维数组。

MyArray5.flatten()

Out [52]: array([1, 6, 11, 16, 2, 7, 12, 17, 3, 8, 13, 18, 4, 9, 14, 19, 5,
      10, 15, 20])

In [53]:

 (5)用 tolist()方法将多维数组转换为嵌套列表。

MyArray5.tolist()

Out [53]: [[1, 6, 11, 16],
  [2, 7, 12, 17],
  [3, 8, 13, 18],
  [4, 9, 14, 19],
  [5, 10, 15, 20]]

36 数组

In [54]:

（6）可以重设数组元素的数据类型。

MyArray5.astype(np.float)

Out [54]: array([[1., 6., 11., 16.],
　　　　　　 [2., 7., 12., 17.],
　　　　　　 [3., 8., 13., 18.],
　　　　　　 [4., 9., 14., 19.],
　　　　　　 [5., 10., 15., 20.]])

In [55]: MyArray5

MyArray5.astype(np.float)执行后，数组MyArray5本身没有变，而是返回另一个新数组。

Out [55]: array([[1, 6, 11, 16],
　　　　　　 [2, 7, 12, 17],
　　　　　　 [3, 8, 13, 18],
　　　　　　 [4, 9, 14, 19],
　　　　　　 [5, 10, 15, 20]])

36.6 属性计算

In [56]:

（1）计算数组的秩——rank()或ndim()。

np.rank(MyArray5)

系统提示"'rank' is deprecated"，说明这个方法已经淘汰，系统显示改用ndim "use the 'ndim' attribute or function instead"。Python第三方包中这种命名方法的变化较为常见。

C:\Anaconda\lib\site-packages\ipykernel_launcher.py:3: Visible-Deprecatio nWarning: 'rank' is deprecated; use the 'ndim' attribute or function ins tead. To find the rank of a matrix see 'numpy.linalg.matrix_rank'.
　This is separate from the ipykernel package so we can avoid doing imports until

Out [56]: 2

In [57]: `np.ndim(MyArray5)`

Out [57]: 2

In [58]: `MyArray5.ndim`

> 提示 ndim 的用法有两种：属性形式，MyArray5.ndim；方法形式，np.ndim(MyArray5)。

Out [58]: 2

In [59]:
> 提示 （2）数组的形状——shape()方法或 shape 属性。

`np.shape(MyArray5)`

Out [59]: (5, 4)

In [60]: `MyArray5.shape`

> 提示 shape 属性支持函数式调用，如 np.shape(MyArray4)等同于 MyArray4.shape。

Out [60]: (5, 4)

In [61]:
> 提示 （3）计算元素个数——.size。

`MyArray5.size`

> 提示 数组有三个常用属性：shape、ndim、size。

Out [61]: 20

In [62]:
> 提示 （4）查看数组类型——内置函数 type()。

`type(MyArray5)`

36 数组

 注意 type 不是 NumPy 提供的,而是 Python 内置函数,所以不能加前缀 np。

Out [62]: numpy.ndarray

36.7 ndarray 的计算

In [63]:
 提示 (1)数组的乘法。

MyArray5*10

Out [63]: array([[10, 60, 110, 160],
 [20, 70, 120, 170],
 [30, 80, 130, 180],
 [40, 90, 140, 190],
 [50, 100, 150, 200]])

In [64]:
 提示 (2)横向拆分——split()方法。

x = np.array([11,12,13,14,15,16,17,18])
x1,x2,x3 = np.split(x,[3,5])

 注意 [3,5]为拆分位置的索引。

print(x1,x2,x3)

Out [64]: [11 12 13] [14 15] [16 17 18]

In [65]:
 提示 纵向拆分——vsplit()方法,此处为元组的拆包式赋值。

upper,lower = np.vsplit(MyArray5.reshape(4,5),[2])
print("上半部分为\n",upper)
print("\n\n 下半部分为\n",lower)

Out [65]: 上半部分为
 [[1 6 11 16 2] [7 12 17 3 8]]

下半部分为
[[13 18 4 9 14] [19 5 10 15 20]]

In [66]:

 （3）数组的合并——np.concatenate()。

np.concatenate((lower, upper), axis = 0)

 axis = 0 代表的是第 0 轴，其本质含义如下。

#1. 计算之后列数不变
#2. 以列为单位进行计算
#3. 逐列计算

Out [66]: array([[13, 18, 4, 9, 14],
　　　　　　　　[19, 5, 10, 15, 20],
　　　　　　　　[1, 6, 11, 16, 2],
　　　　　　　　[7, 12, 17, 3, 8]])

In [67]:

 （4）np.vstack()和 np.hstack()分别支持横向或纵向合并。

np.vstack([upper, lower])

 调用 np.vstack()的前提——列的个数一样。

Out [67]: array([[1, 6, 11, 16, 2],
　　　　　　　　[7, 12, 17, 3, 8],
　　　　　　　　[13, 18, 4, 9, 14],
　　　　　　　　[19, 5, 10, 15, 20]])

In [68]: np.hstack([upper, lower])

 调用 np.hstack()的前提——行的个数一样。

Out [68]: array([[1, 6, 11, 16, 2, 13, 18, 4, 9, 14],
　　　　　　　　[7, 12, 17, 3, 8, 19, 5, 10, 15, 20]])

36 数组

In [69]:
> **提示** 在 NumPy 中，数组的函数计算往往为"ufunc 类函数"——以列为单位计算的函数，目的是支持向量计算，不用写循环语句。

```python
np.add(MyArray5,1)
```

> **注意** 同一个功能，用 sum() 和 np.sum() 都可以实现，但是：

```
#前者为 Python 内置函数，后者为 numpy 包的函数，属于 ufunc
#ufunc 的速度比普通函数快
#sum 为内置函数，并不是 ufunc
#np.sum 为 ufunc 函数。
```

Out [69]:
```
array([[ 2,  7, 12, 17],
       [ 3,  8, 13, 18],
       [ 4,  9, 14, 19],
       [ 5, 10, 15, 20],
       [ 6, 11, 16, 21]])
```

36.8 ndarray 的元素类型

In [70]:
> **注意** ndarray 中的数组支持用户自定义元素的类型，定义方法——设置 dtype 参数值。

```python
import numpy as np
np.zeros(10, dtype = "int16")
```

Out [70]: array([0, 0, 0, 0, 0, 0, 0, 0, 0, 0], dtype = int16)

In [71]:
```python
np.zeros(10, dtype = "float")
```

Out [71]: array([0., 0., 0., 0., 0., 0., 0., 0., 0., 0.])

In [72]:
> **注意** 同一个数组中，所有元素的数据类型必须一致。

```
#如果不一致或没有显式定义元素数据类型，则按 dtype = object 来处理
```

```
a1 = np.array([1,2,3,None])
a1
```

Out [72]: array([1, 2, 3, None], dtype = object)

In [73]:
```
a1 = np.array([1,2,3,None,np.nan])
a1
```

Out [73]: array([1, 2, 3, None, nan], dtype = object)

36.9 插入与删除

In [74]:
```
import numpy as np
myArray1 = np.array([11,12,13,14,15,16,17,18])
np.delete(myArray1,2)
```

 提示 | 删除特定元素的方法——np.delete()。

Out [74]: array([11, 12, 14, 15, 16, 17, 18])

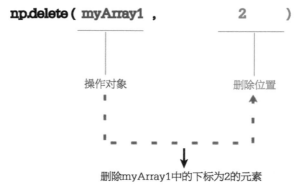

In [75]:
```
np.insert(myArray1,1,88)
```

 提示 | 插入特定元素方法——np.insert()。

Out [75]: array([11, 88, 12, 13, 14, 15, 16, 17, 18])

36.10 缺失值处理

In [76]:

 提示　判断数组的每个元素是否为缺失值——np.isnan()。

np.isnan(myArray)

Out [76]: array([False, False, False, False, False, False, False, False, False, False])

In [77]:

 提示　判断数组中是否至少有一个缺失值——np.any()。

np.any(np.isnan(myArray))

Out [77]: False

In [78]:

 提示　判断数组中的所有元素是否为缺失值——np.all()。

np.all(np.isnan(myArray))

Out [78]: False

In [79]:

 思路　缺失值的处理。

#在很多函数计算中，如遇到缺失值，会报错或得到 NaN 值
#处理方法是用 NaN-Safty 函数，如 np.nansum()

MyArray = np.array([1,2,3,np.nan])

 注意　np.nan 与 None 的区别。

#None 是 Python 基础语法提供的特殊数据类型，不能参加算术运算
#np.nan 是 numpy 提供的数据类型，属于 float 类型，可以参加算术运算

np.nansum(MyArray)

Out [79]: 6.0

In [80]:
```
np.sum(MyArray)
```
> 注意：在 NumPy 中，np.nan 是 float 类型，可以参加算术运算。

Out [80]: nan

36.11 ndarray 的广播规则

In [81]:
> 提示：规则一：如果列数一样但行数不一样，那么进行以行为单位的广播操作，进行循环补齐。
```
import numpy as np
A1 = np.array(range(1,10)).reshape([3,3])
A1
```

Out [81]: array([[1, 2, 3],
         [4, 5, 6],
         [7, 8, 9]])

In [82]:
```
A2 = np.array([10,10,10])
A2
```
> 提示：A1 和 A2 的列数一样，行数不同。

Out [82]: array([10, 10, 10])

In [83]:
```
A1+A2
```
> 提示：在 A1+A2 操作之前，进行以行为单位的广播操作，将 A1 和 A2 转换为相同结构后进行计算。

Out [83]: array([[11, 12, 13],
         [14, 15, 16],
         [17, 18, 19]])

36 数组

In [84]:
> 提示 规则二：如果列数不一致（除了列数为1），则解释器将报错。
>
> A3 = np.arange(10).reshape(2,5)
>
> A3

Out [84]: array([[0, 1, 2, 3, 4],
         [5, 6, 7, 8, 9]])

In [85]:
> A4 = np.arange(16).reshape(4,4)
> A4
>
> 提示 A3为2行×5列，A4为4行×4列。

Out [85]: array([[0, 1, 2, 3],
         [4, 5, 6, 7],
         [8, 9, 10, 11],
         [12, 13, 14, 15]])

In [86]:
> A3+A4
> #报错：ValueError: operands could not be broadcast together with shapes (2,5) (4,4)

Out [86]: ---------------------------------------
ValueError   Traceback (most recent call last)
<ipython-input-86-0fe8480883de>in <module>()
----> 1 A3+A4
      2 #报错：ValueError: operands could not be broadcast together with shapes (2,5) (4,4)

ValueError: operands could not be broadcast together with shapes (2,5) (4,4)

36.12 ndarray 的排序

In [87]:
> import numpy as np
> myArray = np.array([11,18,13,12,19,15,14,17,16])
> myArray

Out [87]: array([11, 18, 13, 12, 19, 15, 14, 17, 16])

In [88]:
> （1）返回排序结果——np.sort()。
>
> np.sort(myArray)

Out [88]: array([11, 12, 13, 14, 15, 16, 17, 18, 19])

In [89]:
> （2）返回排序后的 index——np.argsort()。
>
> np.argsort(myArray)

Out [89]: array([0, 3, 2, 6, 5, 8, 7, 1, 4], dtype=int64)

In [90]:
> （3）多维数组按指定维度排序——np.sort()中加一个轴参数 axis。
>
> MyArray = np.array([[21, 22, 23, 24, 25],
>   [35, 34, 33, 32, 31],
>   [1, 2, 3, 100, 4]])

In [91]:
> np.sort(MyArray, axis = 1)
>
> axis = 1 的含义如下。
>
> #(1) 计算前后的行数不变
> #(2) 以行为单位，每行独立计算
> #(3) 逐行计算

Out [91]: array([[21, 22, 23, 24, 25],
                [31, 32, 33, 34, 35],
                [1, 2, 3, 4, 100]])

In [92]:
> np.sort(MyArray, axis = 0)
>
> axis = 0 的含义如下。

36 数组

```
#(1) 计算前后的列数不变
#(2) 以列为单位,每列独立计算
#(3) 逐列计算
```

Out [92]:　array([[1, 2, 3, 24, 4],
       　　　　　[21, 22, 23, 32, 25],
       　　　　　[35, 34, 33, 100, 31]])

> **扩展** 一个致力于在数据科学和大数据领域消除性别歧视的社区:Girls Who Code,其首页如图 36-1 所示。

图 36-1　Girls Who Code 首页

37 Series

 常见疑问及解答。

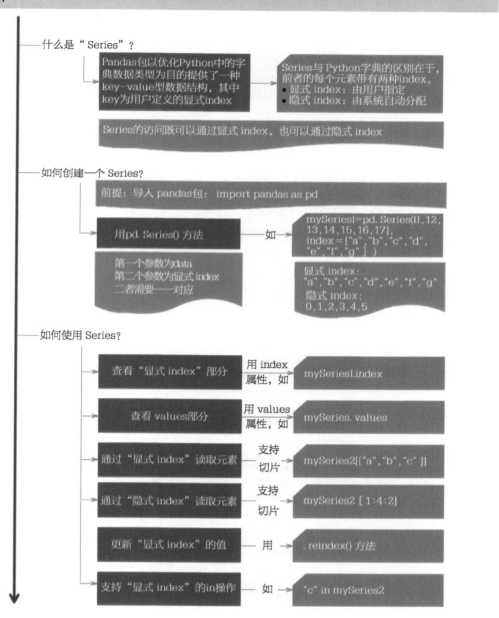

37 Series

37.1 Series 的主要特点

In [1]:

> 特点之一：Series 是一种 key-value 型数据结构，每个元素由如下两部分组成。

```
#key: 多个显式index
#value: 每个显式index对应的value（值）
```

> 特点之二：Series 有两种 index。

```
#显式 index: 定义一个 Series 对象时, 显式地指定 index
#隐式 index: Series 对象中每个元素的下标类似 Python "列表"的下标
```

> 与软件开发不同的是，在数据分析和数据科学项目中，一般用显式索引（index）而不是隐式下标，原因在于：当数据量很大时，很难准确定位其下标。

37.2 Series 的定义方法

In [2]:

 用 pd.Series()函数，函数参数如下。

```
#data参数: 对应的是values
    #取值: 元组、列表等
#index参数: 对应的是显式index
    #取值: 与第一个参数等长的元组、列表等

import pandas as pd
mySeries1 = pd.Series (data = [11,12,13,14,15,16,17], index = ["a","b","c","d","e"," f","g"])
```

 特别注意如下两个问题。

```
#index与values的个数应一致，否则报错
#index为字符串时，别忘了用双引号或单引号括起来，否则报错。常见错误提示
为：NameError: name 'c' is not defined
mySeries1
```

Out[2]:
```
a    11
b    12
c    13
d    14
e    15
f    16
g    17
dtype: int64
```

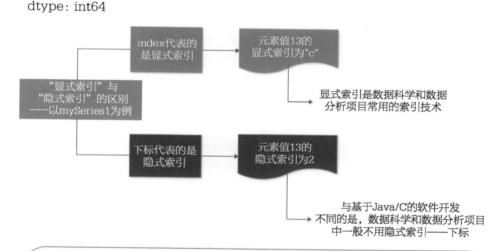

In [3]:
```
import pandas as pd
mySeries1=pd.Series([11,12,13,14,15,16,17], index=[a,b,c,d,e,f,g])
mySeries1
```

 注意 报错信息为 NameError: name 'a' is not defined。原因分析：index 为字符串时，忘记了用双引号或单引号括起来。

Out[3]:
```
---------------------------------
NameError Traceback (most recent call last)
<ipython-input-4-30a072647f76> in <module>()
      1 import pandas as pd
----> 2 mySeries1 = pd.Series([11,12,13,14,15,16,17], index = [a,b,c,d,e,
```

37 Series

```
f,g])
    3
    4 mySeries1

NameError: name 'a' is not defined
```

In [4]:

 在Pandas的早期版本中，当data只包含一个元素时，Series对象的定义支持"循环补齐"。

```
mySeries2 = pd.Series([10], index = ["a","b","c","d","e","f","g"])
mySeries2
```

Out [4]:
```
a    10
b    10
c    10
d    10
e    10
f    10
g    10
dtype: int64
```

In [5]:

 当 data 中的 values 多于一个时，values 和 index 的个数应一致。

```
mySeries3 = pd.Series([1,2,3,4,5], index = ["a","b","c"])
mySeries3
```

 报错信息：ValueError: Wrong number of items passed 5, placement implies 3。
原因分析：values 与 index 的个数不一致。

Out[5]:
```
---------------------------------------------------------------
ValueError                                Traceback (most recent call last)
<ipython-input-20-0785eae88a5c> in <module>()
      1
----> 2 mySeries3 = pd.Series([1,2,3,4,5], index = ["a","b","c"])
      3
      4 mySeries3

C:\Anaconda\lib\site-packages\pandas\core\series.py in __init__(self, data, index, dtype, name, copy, fastpath)
```

```
    264                                    raise_cast_failure=True)
    265
--> 266        data = SingleBlockManager(data, index, fastpath = True)
    267
    268        generic.NDFrame.__init__(self, data, fastpath = True)

C:\Anaconda\lib\site-packages\pandas\core\internals.py in __init__(self, block, axis, do_integrity_check, fastpath)
   4400    if not isinstance(block, Block):
   4401        block = make_block(block, placement=slice(0, len(axis)), ndim = 1,
-> 4402                                       fastpath = True)
   4403
   4404    self.blocks = [block]

C:\Anaconda\lib\site-packages\pandas\core\internals.py in make_block(values, placement, klass, ndim, dtype, fastpath)
   2955                                   placement=placement, dtype = dtype)
   2956
-> 2957        return klass(values, ndim = ndim, fastpath = fastpath, placement = placement)
   2958
   2959 # TODO: flexible with index = None and/or items = None

C:\Anaconda\lib\site-packages\pandas\core\internals.py in __init__(self, values, placement, ndim, fastpath)
    118            raise ValueError('Wrong number of items passed %d, placement '
    119                             'implies %d' % (len(self.values),
--> 120                                             len(self.mgr_locs)))
    121
    122    @property

ValueError: Wrong number of items passed 5, placement implies 3
```

37.3 Series 的操作方法

In [6]:

提示

（1）查看显式 index 部分。

37 Series

```
import pandas as pd
mySeries4 = pd.Series([21,22,23,24,25,26,27], index = ["a","b","c","d","e","f","g"])
mySeries4.index
```

 返回的数据类型为 index，是 Pandas 中定义的一个特殊数据类型。

Out [6]: Index(['a', 'b', 'c', 'd', 'e', 'f', 'g'], dtype = 'object')

In [7]:

 （2）查看 values 部分。

```
mySeries4.values    #此处 values 的拼写方法为复数
```

Out [7]: array([21, 22, 23, 24, 25, 26, 27], dtype = int64)

In [8]:

 （3）可以通过显式 index 查看元素，支持切片，详见本书【14 列表（P76）】。

 key 必须带引号，单引号和双引号都可以。

```
mySeries4['b']
```

Out [8]: 22

In [9]:
```
mySeries4["b"]
```

Out [9]: 22

In [10]:
```
mySeries4[["a","b","c"]]
```

 显式 index 中支持切片。此处有两个方括号"[[]]"，分别为 Series 和切片的"[]"。

Out [10]: a   21
b   22

```
            c    23
        dtype: int64
```

In [11]:
```
mySeries4["a":"d"]
```

 提示 | 显式 index 可以作为 start 和 stop 位置。

 注意 | 除了常量，显式 index 必须用双引号或单引号括起来。

Out[11]:
```
a    21
b    22
c    23
d    24
dtype: int64
```

In [12]:

 提示 | （4）支持通过隐式 index 读取元素。

```
mySeries4[1:4:2]
```

Out[12]:
```
b    22
d    24
dtype: int64
```

In [13]:
```
mySeries4
```

Out[13]:
```
a    21
b    22
c    23
d    24
e    25
f    26
g    27
dtype: int64
```

In [14]:

 提示 | （5）支持显式 index 的 in 操作。其中，Series 的 in 操作以 key 为准。

```
"c" in mySeries4
```

37 Series

Out [14]: True

In [15]: `"h" in mySeries4`

Out [15]: False

In [16]:

（6）更新显式 index 的方法：.reindex()

```
import pandas as pd
mySeries4 = pd.Series([21,22,23,24,25,26,27], index = ["a","b","c","d","e","f","g"])
mySeries5 = mySeries4.reindex(index = ["b","c","a","d","e","g","f"])
mySeries5
```

查看reindex()之后的结果。其中，reindex()修改的是隐式索引而不是显式索引，即更改的是key或value的显示顺序，而不会破坏key和value之间的对应关系。

Out [16]:
```
b    22
c    23
a    21
d    24
e    25
g    27
f    26
dtype: int64
```

In [17]:
```
mySeries5 = mySeries4.reindex(index = ["b","c","a","d","e","g","f"])
mySeries4
```

从输出结果看，在 reindex 的过程中，mySeries4 本身未改变。

Out [17]:
```
a    21
b    22
c    23
d    24
e    25
f    26
g    27
```

```
dtype: int64
```

In [18]:
```
mySeries5 = mySeries4.reindex(index = ["new1","c","a","new2","e",
"g","new3"])
```

 当调用reindex()方法。若提供未知列名，NumPy将在原Seiries对象中自动增加key（显式index），对应的value为NaN缺失值。

```
mySeries5
```

Out [18]:
```
new1    NaN
c       23.0
a       21.0
new2    NaN
e       25.0
g       27.0
new3    NaN
dtype: float64
```

In [19]:
```
mySeries4
```

 reindex()不改变Series对象本身的显式index。

Out [19]:
```
a    21
b    22
c    23
d    24
e    25
f    26
g    27
dtype: int64
```

38 DataFrame

常见疑问及解答。

— 什么是 DataFrame (数据框)?
- Pandas包提供的一种类似关系表的数据结构
- DataFrame代表的是一种类似关系表类的数据结构,是R和Python编程中常用的数据结构之一
- DataFrame是数据科学中最为广泛应用的数据结构之一

— 如何创建一个 DataFrame?
- 前提:导入 pandas包,导入方法为: import pandas as pd
- 直接定义方法 → pd.DataFrame()方法
- 导入外部文件方法 → 用 Pandas包导入一个外部文件时,自动将其转换为DataFrame对象

— DataFrame的数据结构是怎样的?

	名称	个数	显式索引
行	index	index.size shape[0]	index
列	columns	columns.size shape[1]	columns

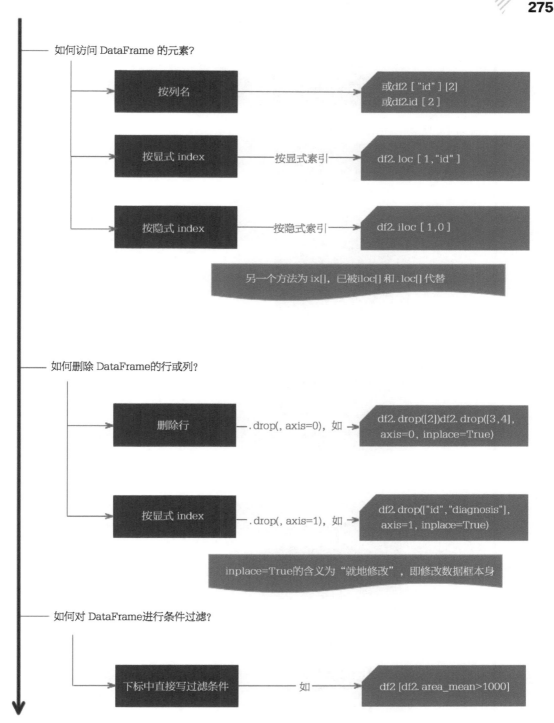

38 DataFrame

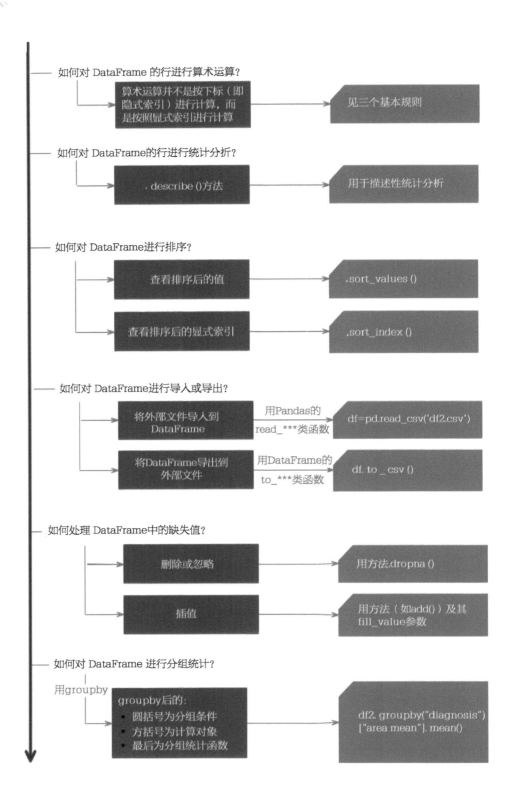

38.1 DataFrame 的创建方法

In [1]:
> **思路** 在数据分析和数据科学项目中,创建一个DataFrame的常用方法有如下两种。
>
> #第一种:直接定义(很少用)
> #第二种:导入定义(很常用)

In [2]:
> **提示** (1)直接定义——pd.DataFrame()。
>
> **注意** pd.DataFrame()的参数可以是 ndarray、列表、字典、元组、Series等。
>
> ```
> import numpy as np
> import pandas as pd
> df1 = pd.DataFrame(np.arange(10).reshape(2,5))
> df1
> ```

Out [2]:

	0	1	2	3	4
0	0	1	2	3	4
1	5	6	7	8	9

In [3]:
> **提示** (2)导入定义:当用 Pandas 包导入一个外部文件时,将自动转换为 DataFrame 对象。
>
> ```
> df2 = pd.read_csv('bc_data.csv')
> df2.shape
> ```
>
> **提示** 有时导入外部文件时出现"乱码"现象,原因在于字符集编码有误,纠正方法为设置属性 encoding。

Out [3]: (569, 32)

In [4]:
```
df2 = df2[["id","diagnosis","area_mean"]]
```

38 DataFrame

在此使用 Fancy Indexing 方法，选择（投影）数据框 df2 的 id、diagnosis、area_mean 等 3 列，参见本书【36 数组(P245)】中对 Fancy Indexing 的解释。

df2.head()

head()和tail()是数据分析与数据科学项目中常用的两个函数，分别用于显示数据框的前几行和后几行。当数据量很大时，无法也没有必要显示全部内容。

Out [4]:

	id	diagnosis	area_mean
0	842302	M	1001.0
1	842517	M	1326.0
2	84300903	M	1203.0
3	84348301	M	386.1
4	84358402	M	1297.0

38.2 查看行或列

In [5]:

查看行名，即行的显式索引——用 index 属性。

df2.index

Out [5]: RangeIndex(start = 0, stop = 569, step = 1)

In [6]:

计算行数——用 index.size 属性。

df2.index.size

Out [6]: 569

In [7]:

查看列名，即列的显式索引——用 columns 属性。

df2.columns

Out [7]: Index(['id', 'diagnosis', 'area_mean'], dtype = 'object')

In [8]:
 计算列数——用 columns.size 属性。

df2.columns.size

Out [8]: 3

In [9]:
 同时显示行数和列数，即查看 DataFrame 的形状，用 shape 属性。

df2.shape

Out [9]: (569, 3)

In [10]:
 计算行数和列数的另一种方法——shape[0]和 shape[1]。

print("行数为:", df2.shape[0])
print("列数为:", df2.shape[1])

> 注意：从 In[9]可看出，"df2.shape" 的返回值为一个元组(569, 3)。因此，df2.shape[0]和 df2.shape[1]分别为该元组的第 0 个元素和第 1 个元素。

Out [10]: 行数为: 569
列数为: 3

38.3 引用行或列

In [11]:
 Python 中数据框的下标的写法很特殊，不能像 C 和 Java 那样写成 df2[1][2]，也不能像 R 语言那样写成 df2[1,2]。

 在 Python 中，可以通过 iloc 达到 "类似 R 语言的下标（即隐式索引）表示方法"，如 df2.iloc[1,0]。

38 DataFrame

In [12]:
> 提示 （1）按列名读取。

In [13]:
```
#第一种写法——列名出现在下标中
df2["id"].head()
```

Out [13]:
```
0       842302
1       842517
2     84300903
3     84348301
4     84358402
Name: id, dtype: int64
```

In [14]:
```
#第二种写法——可将列名当作数据框的一个属性来用
df2.id.head()
```

Out [14]:
```
0       842302
1       842517
2     84300903
3     84348301
4     84358402
Name: id, dtype: int64
```

In [15]:
```
#第三种写法——列名和行号一起用
df2["id"][2]
```

 注意 数据框的第 0 轴为列，所以不能写成 df2[2]["id"]，否则报错——KeyError: 2。

Out [15]: 84300903

In [16]:
```
#第四种写法——属性名和行号一起用
df2.id[2]
```

Out [16]: 84300903

In [17]:
```
#第五种写法——用切片
df2["id"][[2,4]]
```

Out [17]:
```
2    84300903
4    84358402
Name: id, dtype: int64
```

In [18]:

提示 （2）按index。

注意 在Pandas中，每个数据框有两种index（索引），一种是自带或默认的，从0开始；另一种是通过index属性定义的。

```
#前者被称为隐式index，后者被称为显式index
#显式index：由用户自行定义，当然也可能定义成0、1、2、3这样的整数。若index设为整数，可能导致混乱，到底是哪一种索引？
#为此，引入了loc、iloc、ix等属性
    #loc表示的是显式index
    #iloc 表示的是隐式 index，即implicit index。
```

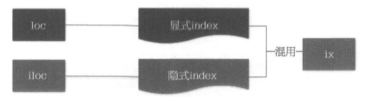

思路 与C和Java不同的是，Python中的DataFrame计算的默认（或优先）依据并非位置（或隐式index），而是显式index。

```
df2.loc[1,"id"]
```

注意 loc、iloc、ix 后面是方括号，而不是圆括号。其中，ix 为 Pandas 的早期版本中支持的一种写法，从 Pandas 0.20.0 开始已取消此种写法。

Out [18]: 842517

In [19]:

提示 （3）按位置，即按隐式index。

38 DataFrame

```
df2.iloc[1,0]
```

Out [19]: 842517

In [20]:

 提示 | 按显式 index 访问非连续元素。

```
df2[["area_mean","id"]].head()
```

 注意 | 如果在下标中出现多个显式 index，则必须用切片方法，参见本书【14 列表（P80/In[8]）】。

Out [20]:

	area_mean	id
0	1001.0	842302
1	1326.0	842517
2	1203.0	84300903
3	386.1	84348301
4	1297.0	84358402

38.4 index 操作

In [21]:

 提示 | （1）DataFrame 的行和列都有自己的名称，即显式 index。

```
#行的显式index的名称：index
#列的显式index的名称：columns
df2.index
```

 技巧 | df2.index 的返回值为惰性计算的迭代器 RangeIndex()。我们可以用 print(*df2.index)方式直接显示/打印其具体值。

Out [21]: RangeIndex(start = 0, stop = 569, step = 1)

In [22]:
```
df2.columns
```

Out [22]: Index(['id', 'diagnosis', 'area_mean'], dtype = 'object')

In [23]:

 提示 （2）可以按显式 index 读取。

df2["id"].head()

Out [23]:
```
0        842302
1        842517
2      84300903
3      84348301
4      84358402
Name: id, dtype: int64
```

In [24]:

 提示 （3）更改显式 index 的方法：用 reindex()方法。

df2.reindex(index = ["1","2","3"],columns = ["1","2","3"])
df2.head()

 注意 与 Series 类似，DataFrame 的 reindex()方法更改的是隐式 index，即显示顺序，而不会破坏显式 index 与数据内容的对应关系。

Out [24]:

	id	diagnosis	area_mean
0	842302	M	1001.0
1	842517	M	1326.0
2	84300903	M	1203.0
3	84348301	M	386.1
4	84358402	M	1297.0

In [25]: df2.reindex(index = [2,3,1], columns = ["diagnosis","id","area_mean"])

Out [25]:

	diagnosis	id	area_mean
2	M	84300903	1203.0
3	M	84348301	386.1
1	M	842517	1326.0

In [26]:

 提示 在重新索引时可以新增一个显式 index。

38 DataFrame

```
df3 = df2.reindex(index = [2,3,1], columns = ["diagnosis","id","area_mean","MyNewColumn"],fill_value=100)
```

 注意 此处，新增 index 的名称为 MyNewColumn。

```
df3
```

Out [26]:

	diagnosis	id	area_mean	MyNewColumn
2	M	84300903	1203.0	100
3	M	84348301	386.1	100
1	M	842517	1326.0	100

38.5 删除或过滤行/列

In [27]:
```
import pandas as pd
df2 = pd.read_csv('bc_data.csv')
```

 提示 用 Pandas 的 read_csv() 方法将外部文件 bc_data.csv 读入本地数据框 df2。

```
df2 = df2[["id","diagnosis","area_mean"]]
df2.head()
```

Out [27]:

	id	diagnosis	area_mean
0	842302	M	1001.0
1	842517	M	1326.0
2	84300903	M	1203.0
3	84348301	M	386.1
4	84358402	M	1297.0

In [28]:
```
df2.drop([2]).head()
```

 提示 此处，下标中的 2 为显式 index，而不是隐式 index。

Out [28]:

	id	diagnosis	area_mean
0	842302	M	1001.0
1	842517	M	1326.0
3	84348301	M	386.1
4	84358402	M	1297.0
5	843786	M	477.1

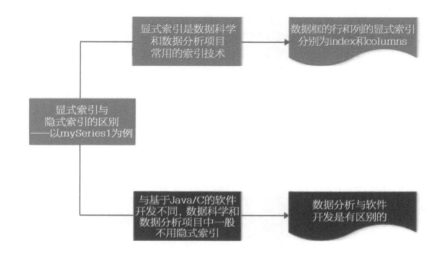

In [29]:
```
df2.head()
```

 提示　drop()方法不修改 DataFrame 对象本身。

Out [29]:

	id	diagnosis	area_mean
0	842302	M	1001.0
1	842517	M	1326.0
2	84300903	M	1203.0
3	84348301	M	386.1
4	84358402	M	1297.0

In [30]:
```
import pandas as pd
df2 = pd.read_csv('bc_data.csv')
df2 = df2[["id","diagnosis","area_mean"]]
```

38 DataFrame

```
df2.drop([3,4], axis = 0, inplace = True)
```

 如果前两行代码没有写，那么反复运行此行代码会报错。
原因：df2 的当前值一直在发生变化。

 axis = 0 的含义如下。

#（1）计算前后的列数不变
#（2）以列为单位计算
#（3）逐列计算

 df2.drop()的第一个参数可以为列表，也可以为一个值。

```
df2.head()
```

Out [30]:

	id	diagnosis	area_mean
0	842302	M	1001.0
1	842517	M	1326.0
2	84300903	M	1203.0
5	843786	M	477.1
6	844359	M	1040.0

In [31]:

 Python 中的一个重要参数——inplace。
功能：是否要修改 DataFrame 对象本身（就地修改）。

#取值
　#inplace=True，就地修改，即修改DataFrame对象（如df2）本身
　#inplace=False，不修改 DataFrame 对象本身（如 df2），而返回另一个
　　DataFrame 对象

	是否修改数据本身（就地修改）	是否返回一个新的值（返回新值）
inplace=True	是	否
inplace=False	否	是

```
import pandas as pd
df2 = pd.read_csv('bc_data.csv')
df2 = df2[["id","diagnosis","area_mean"]]
df2.drop([3,4], axis = 0, inplace = False)
df2.head()
```

Out [31]:

	id	diagnosis	area_mean
0	842302	M	1001.0
1	842517	M	1326.0
2	84300903	M	1203.0
3	84348301	M	386.1
4	84358402	M	1297.0

In [32]:

 提示 | 删除列的第一个方法——用切片操作和 del 语句。

```
import pandas as pd
df2 = pd.read_csv('bc_data.csv')
df2 = df2[["id","diagnosis","area_mean"]]
del df2["area_mean"]
df2.head()
```

Out [32]:

	id	diagnosis
0	842302	M
1	842517	M
2	84300903	M
3	84348301	M
4	84358402	M

In [33]:

 提示 | 删除列的第二种方法——用 drop()方法。

```
df2.drop(["id","diagnosis"], axis = 1, inplace = True)
df2.head()
```

38 DataFrame

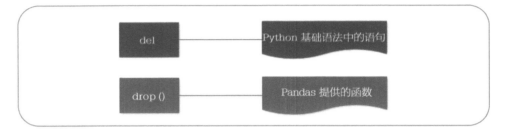

Out [33]:

	area_mean
0	1001.0
1	1326.0
2	1203.0
3	386.1
4	1297.0

In [34]:

 提示 按列条件过滤方法——在"下标"中写过滤条件。

```
import pandas as pd
df2 = pd.read_csv('bc_data.csv')

df2 = df2[["id","diagnosis","area_mean"]]
df2[df2.area_mean > 1000].head()
```

 注意 此处,过滤条件"df2.area_mean> 1000"写在 df2[]的下标中。

Out [34]:

	id	diagnosis	area_mean
0	842302	M	1001.0
1	842517	M	1326.0
2	84300903	M	1203.0
4	84358402	M	1297.0
6	844359	M	1040.0

In [35]:

```
df2[df2.area_mean > 1000][["id","diagnosis"]].head()
```

 提示 此代码为过滤条件"df2.area_mean > 1000"和"切片(["id", "diagnosis"])"的综合运用。

Out [35]:

	id	diagnosis
0	842302	M
1	842517	M
2	84300903	M
4	84358402	M
6	844359	M

38.6 算术运算

In [36]:

> **提示** 规则之一：数据框之间的计算规则——先补齐显式 index（新增索引对应值为 NaN），得到相同结构后，再进行计算。
>
> df4 = pd.DataFrame(np.arange(6).reshape(2,3))
> df4
>
> **注意** 与 C 和 Java 不同的是，Python 中的 DataFrame 计算的依据不是位置（或隐式 index），而是显式 index。

Out [36]:

	0	1	2
0	0	1	2
1	3	4	5

In [37]:

df5 = pd.DataFrame(np.arange(10).reshape(2,5))

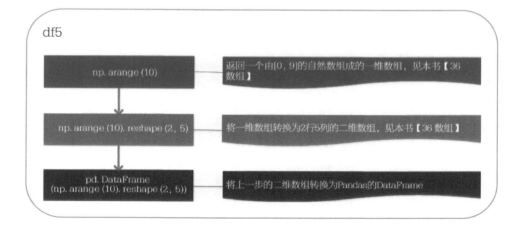

38 DataFrame

Out [37]:

	0	1	2	3	4
0	0	1	2	3	4
1	5	6	7	8	9

In [38]:
```
df4+df5
```

Out [38]:

	0	1	2	3	4
0	0	2	4	NaN	NaN
1	8	10	12	NaN	NaN

In [39]:

 规则之二：用算术运算符+、-、*等会产生 NaN，如果想将默认填充的 NaN 改为指定值，建议不要使用算术运算符，而改用成员方法，如 add()、sub()、mul()、div()。

```
df6 = df4.add(df5, fill_value = 10)
df6
```

 在数据分析或数据科学项目中一般不用运算符，而用对应的成员方法，原因在于：调用成员方法的灵活性比运算符高，成员方法中可以设置更多参数，如缺失值处理参数（fillna）、计算方向（axis）等。

Out [39]:

	0	1	2	3	4
0	0	2	4	13.0	14.0
1	8	10	12	18.0	19.0

In [40]:

 规则之三：数据框与 Series 的计算规则——按行（第 1 轴）广播，先把行改为等长，行内不做循环补齐。只是一行一行计算，不会跨行广播，参见本书【36.11 ndarray 的广播规则 (P261/In[81])】。

```
s1 = pd.Series(np.arange(3))
s1
```

Out [40]: 0   0
1   1

```
2    2
dtype: int32
```

In [41]:
```
df6-s1
```

Out [41]:

	0	1	2	3	4
0	0.0	1.0	2.0	NaN	NaN
1	8.0	9.0	10.0	NaN	NaN

In [42]:
```
df5 = pd.DataFrame(np.arange(10).reshape(2,5))
s1 = pd.Series(np.arange(3))
df5-s1
```

 提示 计算思路为：先对齐显式 index，然后按显式 index 进行计算。

Out [42]:

	0	1	2	3	4
0	0.0	0.0	0.0	NaN	NaN
1	5.0	5.0	5.0	NaN	NaN

In [43]:

 提示 等价于成员方法 sub()中的 axis 参数取值为"1"。四则运算对应的方法分别为 add()、sub()、mul()、div()。

```
df5 = pd.DataFrame(np.arange(10).reshape(2,5))
s1 = pd.Series(np.arange(3))
df5.sub(s1,axis = 1)
```

 注意 axis = 1 的含义可以理解为以下 3 个条件同时满足。

```
#（1）计算前后的行数不变
#（2）以行为单位进行计算
#（3）逐行计算
```

Out [43]:

	0	1	2	3	4
0	0.0	0.0	0.0	NaN	NaN

38 DataFrame

| 1 | 5.0 | 5.0 | 5.0 | NaN | NaN |

In [44]:

 改成按列（第 0 轴）计算的方法：用成员方法/函数，而不用运算符。

 计算规则：先对齐显式 index（把列做成等长），再以显式 index 为依据进行计算。

```
df5 = pd.DataFrame(np.arange(10).reshape(2,5))
s1 = pd.Series(np.arange(3))
df5.sub(s1,axis = 0)
```

Out [44]:

	0	1	2	3	4
0	0.0	1.0	2.0	3.0	4.0
1	4.0	5.0	6.0	7.0	8.0
2	NaN	NaN	NaN	NaN	NaN

In [45]:

```
df7 = pd.DataFrame(np.arange(20).reshape(4,5))
df7
```

Out [45]:

	0	1	2	3	4
0	0	1	2	3	4
1	5	6	7	8	9
2	10	11	12	13	14
3	15	16	17	18	19

In [46]:

```
df7+2
```

Out [46]:

	0	1	2	3	4
0	2	3	4	5	6
1	7	8	9	10	11
2	12	13	14	15	16
3	17	18	19	20	21

In [47]:

 此外,Pandas 提供了更多的函数,以支持不同的处理需求。

```
print(df7)
print("df7.cumsum = ", df7.cumsum())
```

 按列计算时,不会进行跨列计算。

Out [47]:
```
    0   1   2   3   4
0   0   1   2   3   4
1   5   6   7   8   9
2  10  11  12  13  14
3  15  16  17  18  19
df7.cumsum =     0   1   2   3   4
0   0   1   2   3   4
1   5   7   9  11  13
2  15  18  21  24  27
3  30  34  38  42  46
```

In [48]:

```
df7
```

Out [48]:

	0	1	2	3	4
0	0	1	2	3	4
1	5	6	7	8	9
2	10	11	12	13	14
3	15	16	17	18	19

In [49]:

```
df7.rolling(2).sum()
```

 依次计算相邻 2 个元素之和,即本元素与上一个元素之和。

38 DataFrame

 默认情况下，按列计算，即 axis = 0。

Out [49]:

	0	1	2	3	4
0	NaN	NaN	NaN	NaN	NaN
1	5.0	7.0	9.0	11.0	13.0
2	15.0	17.0	19.0	21.0	23.0
3	25.0	27.0	29.0	31.0	33.0

In [50]: df7.rolling(2, axis = 1).sum()

Out [50]:

	0	1	2	3	4
0	NaN	1.0	3.0	5.0	7.0
1	NaN	11.0	13.0	15.0	17.0
2	NaN	21.0	23.0	25.0	27.0
3	NaN	31.0	33.0	35.0	37.0

In [51]: df7.cov()

 协方差矩：.cov()方法阵。更多函数及其用法，请参见 Pandas 官网中的函数介绍文档 https://pandas.pydata.org/pandas-docs/stable/reference/frame.html。

Out [51]:

	0	1	2	3	4
0	41.666667	41.666667	41.666667	41.666667	41.666667
1	41.666667	41.666667	41.666667	41.666667	41.666667
2	41.666667	41.666667	41.666667	41.666667	41.666667
3	41.666667	41.666667	41.666667	41.666667	41.666667
4	41.666667	41.666667	41.666667	41.666667	41.666667

In [52]: df7.corr()

 相关系数矩阵：.corr()方法。

Out [52]:

	0	1	2	3	4
0	1.0	1.0	1.0	1.0	1.0
1	1.0	1.0	1.0	1.0	1.0
2	1.0	1.0	1.0	1.0	1.0
3	1.0	1.0	1.0	1.0	1.0
4	1.0	1.0	1.0	1.0	1.0

In [53]:

 提示 | 数据框的转置：.T 属性。

```
import pandas as pd
df2 = pd.read_csv('bc_data.csv')
df2 = df2[["id","diagnosis","area_mean"]][2:5]
df2.T
```

Out [53]:

	2	3	4
id	84300903	84348301	84358402
diagnosis	M	M	M
area_mean	1203	386.1	1297

38.7 大小比较运算

In [54]:

```
print(df6)
```

Out [54]:

	0	1	2	3	4
0	0	2	4	13.0	14.0
1	8	10	12	18.0	19.0

In [55]:

```
df6>5
```

Out [55]:

	0	1	2	3	4
0	False	False	False	True	True
1	True	True	True	True	True

In [56]:

```
print(s1)
```

38 DataFrame

```
0    0
1    1
2    2
dtype: int32
```

In [57]:
```
df6 > s1
```

Out [57]:

	0	1	2	3	4
0	False	True	True	False	False
1	True	True	True	False	False

In [58]:
```
df6 > (2,18)
```

Out [58]:

	0	1	2	3	4
0	False	False	True	True	True
1	False	False	False	False	True

38.8 统计信息

In [59]:

 提示 方法 describe()是数据分析与数据科学项目中最常用的描述性统计方法之一。除了 describe()方法，还可以调用 info()方法或第三方包 pandas_profiling 进行数据框的描述性统计。

```
import numpy as np
import pandas as pd
df2 = pd.read_csv('bc_data.csv')
df2 = df2[["id","diagnosis","area_mean"]]
df2.describe()
```

Out [59]:

	id	area_mean
count	5.690000e+02	569.000000
mean	3.037183e+07	654.889104
std	1.250206e+08	351.914129
min	8.670000e+03	143.500000
25%	8.692180e+05	420.300000

50%	9.060240e+05	551.100000
75%	8.813129e+06	782.700000
max	9.113205e+08	2501.000000

In [60]:

 提示 数据框的过滤方法：将过滤条件写在"下标"中。

dt = df2[df2.diagnosis == 'M']

In [61]:

 提示 查看前几行数据。

dt.head()

 提示 在数据分析和数据科学项目中，一般数据量都很大，我们没有必要查阅数据的所有内容，而只需看前几行或最后几行即可，因为数据在同一个列上往往是同质的。

Out [61]:

	id	diagnosis	area_mean
0	842302	M	1001.0
1	842517	M	1326.0
2	84300903	M	1203.0
3	84348301	M	386.1
4	84358402	M	1297.0

In [62]:

dt.tail()

 提示 显示最后几行。

Out [62]:

	id	diagnosis	area_mean
563	926125	M	1347.0
564	926424	M	1479.0
565	926682	M	1261.0
566	926954	M	858.1
567	927241	M	1265.0

38 DataFrame

In [63]:

提示 | 频次统计——count()方法。

df2[df2.diagnosis == 'M'].count()

Out [63]:
```
id          212
diagnosis   212
area_mean   212
dtype: int64
```

In [64]:

提示 | 可以采用 Fancy Indexing 来访问非连续的行或列，参见本书【36 数组（P245）】中对 Fancy Indexing 的解释。

df2[["area_mean","id"]].head()

Out [64]:

	area_mean	id
0	1001.0	842302
1	1326.0	842517
2	1203.0	84300903
3	386.1	84348301
4	1297.0	84358402

38.9 排序

In [65]:

提示 | 先查看数据框 df2 的当前值的前 8 行。

df2.head(8)

Out [65]:

	id	diagnosis	area_mean
0	842302	M	1001.0
1	842517	M	1326.0
2	84300903	M	1203.0
3	84348301	M	386.1
4	84358402	M	1297.0

5	843786	M	477.1
6	844359	M	1040.0
7	84458202	M	577.9

In [66]:

 提示 | 按值排序——sort_values()方法。

df2.sort_values(by = "area_mean", axis = 0, ascending = True).head()

Out [66]:

	id	diagnosis	area_mean
101	862722	B	143.5
539	921362	B	170.4
538	921092	B	178.8
568	92751	B	181.0
46	85713702	B	201.9

In [67]:

#按显式 index 排序——sort_index()方法

df2.sort_index(axis = 1).head(3)

 提示 | axis = 1 的含义可以理解为以下三个条件同时成立。

#（1）计算前后的行数不变
#（2）以行为单位进行计算
#（3）逐行计算

Out [67]:

	area_mean	diagnosis	id
0	1001.0	M	842302
1	1326.0	M	842517
2	1203.0	M	84300903

In [68]:

df2.sort_index(axis = 0, ascending = False).head(3)

 提示 | axis = 0 的含义如下，可以理解为以下三个条件同时成立。

#（1）计算前后的列数不变
#（2）以列为单位进行计算
#（3）逐列计算

38 DataFrame

Out [68]:

	id	diagnosis	area_mean
568	92751	B	181.0
567	927241	M	1265.0
566	926954	M	858.1

38.10 导入/导出

In [69]:

 提示 | DataFrame 导入/导出的前提——需要知道当前工作目录的位置，参见本书【34 当前工作目录（P225/In[2]）】。

```
import os
print(os.getcwd())
```

 提示 | 查看当前工作目录的方法——用 os 模块中的 getcwd()。

Out [69]: C:\Users\soloman\clm

In [70]:

 提示 | 写出方法——用系列方法.to_***()，如.to_csv()。

```
df2.head(3).to_csv("df2.csv")
```

In [71]:

 提示 | 读入方法——用系列方法 read_****()，如.read_csv()。

```
import pandas as pd
df3 = pd.read_csv('df2.csv')
```

In [72]:

```
df3
```

Out [72]:

	Unnamed: 0	id	diagnosis	area_mean
0	0	842302	M	1001.0
1	1	842517	M	1326.0
2	2	84300903	M	1203.0

In [73]:
```
#第三种方法——用数据源/数据文件格式对应的包或模块,如csv。
import csv
with open('df2.csv', newline='') as f:
    reader = csv.reader(f)
    for row in reader:
        print(row)
df3 = pd.read_csv('df2.csv')
```

Out [73]: ['', 'id', 'diagnosis', 'area_mean']　　　['0', '842302', 'M', '1001.0']
['1', '842517', 'M', '1326.0']　　　['2', '84300903', 'M', '1203.0']

In [74]:
```
df3
```

Out [74]:

	Unnamed: 0	id	diagnosis	area_mean
0	0	842302	M	1001.0
1	1	842517	M	1326.0
2	2	84300903	M	1203.0

In [75]:

导出方法——to_excel()等。

```
df2.head(3).to_excel("df3.xls")
```

In [76]:

再次导入刚导出的文件df3.xls。

```
df3 = pd.read_excel("df3.xls")
df3
```

Out [76]:

	id	diagnosis	area_mean
0	842302	M	1001
1	842517	M	1326
2	84300903	M	1203

38.11 缺失数据处理

In [77]:

判断一个数据框是否为空数据框——属性.empty。

```
df3.empty
```

38 DataFrame

Out [77]: False

In [78]:
 Python 基础语法与 Pandas 中对 None 和 NaN 的处理方法不同。

```
#在Python基础语法中，None不能参加计算，NaN可以参加计算
#在Pandas中，二者一样，都可以参加计算，将None自动转换为
 np.nan-np.nan +1
```

Out [78]: nan

In [79]:
```
np.nan-np.nan
```

Out [79]: nan

In [80]:
```
None+1
```

 报错信息为 TypeError: unsupported operand type(s) for +: 'NoneType' and 'int'。原因分析：None 不能参加算术运算。

 None 是 Python 基础语法中的特殊数据类型，不属于数值类型，不能参加算术运算。
np.nan 属于 float 类型，可以参加算术运算。

Out [80]:
```
---------------------------------
TypeError                                 Traceback (most recent call last)
<ipython-input-83-6e170940e108>in <module>()
----> 1 None+1

TypeError: unsupported operand type(s) for +: 'NoneType' and 'int'
```

In [81]:
```
import pandas as pd
import numpy as np
A = pd.DataFrame(np.array([10,10,20,20]).reshape(2,2),columns = list("ab"),index = list("SW"))
A
```

Out [81]:

	a	b
S	10	10
W	20	20

In [82]:

 注意 list("ab")的含义为将字符串"ab"强制类型转换为列表,参见本书【21 内置函数(P136/In[11])】中的强制类型转换操作。

```
list("ab")
```

Out [82]: ['a', 'b']

In [83]:

```
B = pd.DataFrame(np.array([1,1,1,2,2,2,3,3,3]).reshape(3,3), columns = list("abc"),index = list("SWT"))
B
```

Out [83]:

	a	b	c
S	1	1	1
W	2	2	2
T	3	3	3

In [84]:

C=A+B

 注意 初学者需要注意以下三个问题。

#计算依据并非元素的对应位置,而是"以行和列的显式 index 为依据"计算
#缺失值用 NaN 表示
#基本计算流程:先补显式 Index 索引,并在新增显式 Index 中自动补 NaN,再按显式 Index 计算
C

Out [84]:

	a	b	c
S	11.0	11.0	NaN
T	NaN	NaN	NaN
W	22.0	22.0	NaN

In [85]:

 提示 在数据分析或数据科学项目中,一般不用运算符(如"+"),而是用成员方法/函数(如 add())。

38 DataFrame

```
#原因：后者更灵活，如设置或调整参数。
A.add(B, fill_value = 0)
```

 fill_value = 0 的含义：缺失值的处理方式为补 NaN，在此改为补 0。

Out [85]:

	a	b	c
S	11.0	11.0	1.0
T	3.0	3.0	3.0
W	22.0	22.0	2.0

In [86]:

```
A.add(B, fill_value = A.stack().mean())
```

 fill_value = A.stack().mean()的含义为，缺失值用"均值"插补。

Out [86]:

	a	b	c
S	11.0	11.0	16.0
T	18.0	18.0	18.0
W	22.0	22.0	17.0

In [87]:

注意：A.mean()是按列计算的，如果想计算一个数据框（DataFrame）中全部值的均值，就需要方法 stack()。

```
A.mean()
```

Out[87]: a   15.0
         b   15.0
         dtype: float64

In [88]:

```
A.stack()
```

 方法 stack()的功能为建立多级索引，是 NumPy 的一个常用方法。

Out [88]: S a   10
              b   10
           W a   20

```
b    20
dtype: int32
```

In [89]: `A.stack().mean()`

Out [89]: 15.0

In [90]: DataFrame 缺失值处理的四个重要函数如下。

```
#isnull()
#notnull()
#dropna()
#fillna()
C
```

 NaN 代表的是缺失值。

Out [90]:

	a	b	c
S	11.0	11.0	NaN
T	NaN	NaN	NaN
W	22.0	22.0	NaN

In [91]: （1）方法 isnull()——判断数据框中的每个元素是否为"空"。

`C.isnull()`

Out [91]:

	a	b	c
S	False	False	True
T	True	True	True
W	False	False	True

In [92]: （2）方法 notnull()——判断数据框中的每个元素是否为"非空"。

`C.notnull()`

38 DataFrame

Out [92]:

	a	b	c
S	True	True	False
T	False	False	False
W	True	True	False

In [93]:

 （3）方法 dropna()——直接删除缺失值。

C.dropna(axis='index')

 当 axis = 'index'时，凡是含有缺失值的行均被删除，更多参见.dropna()的帮助信息。

Out [93]:

	a	b	c

In [94]:

 （4）方法 fillna()——设置缺失值的填补方法。

C.fillna(0)

 此处，用 0 来补充所有的缺失值 NaN，但 C 的本身未发生变化。

Out [94]:

	a	b	c
S	11.0	11.0	0.0
T	0.0	0.0	0.0
W	22.0	22.0	0.0

In [95]:

C.fillna(method = "ffill")

 ffill 的含义为缺失值的处理策略为 foward fill，即向前填充。

Out [95]:

	a	b	c
S	11.0	11.0	NaN
T	11.0	11.0	NaN
W	22.0	22.0	NaN

In [96]: C.fillna(method = "bfill",axis = 1)

 bfill 的含义为缺失值的处理策略为 backward fill，即向后填充。

Out [96]:
	a	b	c
S	11.0	11.0	NaN
T	NaN	NaN	NaN
W	22.0	22.0	NaN

38.12 分组统计

In [97]: 读入外部文件 bc_data.csv 至本地数据框 df2。

```
import pandas as pd
df2 = pd.read_csv('bc_data.csv')
```

 投影（或仅保留）数据框 df2 的 3 个列：id, diagnosis, area_mean。

```
df2 = df2[["id","diagnosis","area_mean"]]
```

 用方法 head() 显示数据框 df2 的前 5 行。

```
df2.head()
```

Out[97]:
	id	diagnosis	area_mean
0	842302	M	1001.0
1	842517	M	1326.0
2	84300903	M	1203.0
3	84348301	M	386.1
4	84358402	M	1297.0

In [98]: groupby() 的用法如下。

38 DataFrame

```
#（1）后接一个圆括号和一个方括号
    #圆括号中为分组条件
    #方括号中为计算对象
# (2) 再接一个方法，如sum()和mean()
df2.groupby("diagnosis")["area_mean"].mean()
```

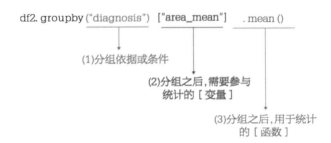

Out[98]: diagnosis
B   462.790196
M   978.376415
Name: area_mean, dtype: float64

In [99]:

> 注意 如果想同时计算多个函数的值，那么需要用方法 aggregate()，将多个函数名以列表形式枚举在方法 aggregate()的参数中。

```
df2.groupby("diagnosis")["area_mean"].aggregate(["mean","sum","max",np.median])
```

Out[99]:

diagnosis	mean	sum	max	median
B	462.790196	165216.1	992.1	458.4
M	978.376415	207415.8	2501.0	932.0

In [100]:

```
df2.groupby("diagnosis")["area_mean"].aggregate(["mean","sum"]).unstack()
```

> 提示 此处，unstack()函数的功能为"将关系表转换为二级索引"。此外，stack()、unstack()、pivot()、melt()是数据分析中常用的数据格式转换函数，建议读者自行了解上述函数的功能。

> **注意** stack()和 unstack()是数据分析或数据科学项目中常用函数,用于建立和撤销多级索引。

Out[100]: diagnosis
   mean B 462.790196
      M 978.376415
   sum B 165216.100000
      M 207415.800000
   dtype: float64

In [101]:

> **提示** apply()方法将第三个位置上的函数替换成自定义函数。更多函数及其用法,请参见 Pandas 官网的函数介绍文档 https://pandas.pydata.org/pandas-docs/stable/reference/frame.html。

```python
def myfunc(x):
    x["area_mean"] /= x["area_mean"].sum()
    return x

df2.groupby("diagnosis").apply(myfunc).head()
```

Out[101]:

	id	diagnosis	area_mean
0	842302	M	0.004826
1	842517	M	0.006393
2	84300903	M	0.005800
3	84348301	M	0.001861
4	84358402	M	0.006253

扩展 数据加工。

  数据加工(Data Wrangling 或 Data Munging)是大数据时代的新关注点之一,不完全等同于传统意义上的数据处理(Data Processing)。二者的主要区别在于:数据加工过程更强调的是数据的转换和价值提升,将数据科学家的 3C 精神融入数据转换过程,追求的是数据处理过程的创新与增值。相关内容参见朝乐门编著的图书《数据科学理论与实践》。

39 日期与时间

常见疑问及解答。

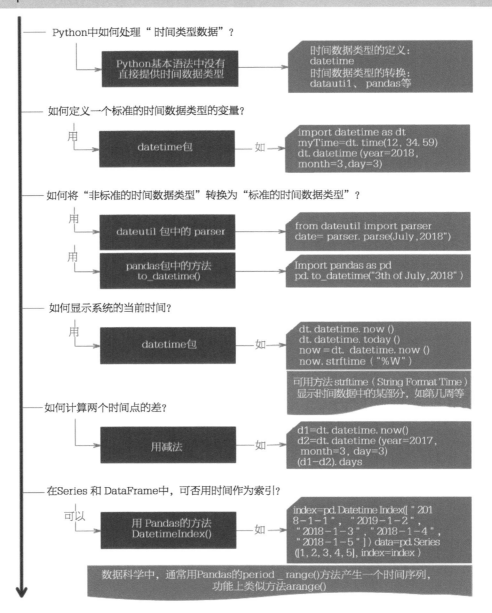

39.1 常用包与模块

In [1]:
```
#日期类型：datetime等
#日期类型的格式转换：dateutil、Pandas 等
```

39.2 时间和日期类型的定义

In [2]:

> **提示** 定义一个标准格式时间类型的对象的方法：用 datetime 包。

```
import datetime as dt
```

> **提示** （1）定义时间类型的数据 dt.time()。

```
myTime = dt.time(12,34,59)
print("myTime:",myTime)
print("myTime.hour:",myTime.hour)
print("myTime.minute:",myTime.minute)
print("myTime.second:",myTime.second)
```

> **技巧** 在 Jupyter Notebook 中，可以输入"myTime."后再按 Tab 键，从而查看系统自动提示，来了解和选择更多的属性。

Out [2]:
myTime: 12:34:59
myTime.hour: 12
myTime.minute: 34
myTime.second: 59

In [3]:

> **提示** （2）定义日期类型的数据：dt.datetime()。
> 需要提醒的是，dt.datetime()和 dt.time()是两个不同函数。

```
dt.datetime(year = 2018, month = 3, day = 3)
```

39 日期与时间

 在方法 datetime()中,year、month、day 为必选,其他为可选。必选参数与可选参数的区别,请参见本书【23 自定义函数(P150/In[12])】。

Out [3]: datetime.datetime(2018, 3, 3, 0, 0)

In [4]:

 也可以通过 "dt.datetime?" 查看帮助信息。

dt.datetime?

 系统显示的帮助信息如下。

```
# Init signature: dt.datetime(self, /, *args, **kwargs)
# Docstring:
# datetime(year, month, day[, hour[, minute[, second[, micro-
second[,tzinfo]]]]])
# The year, month and day arguments are required. tzinfo may be None, or an
# instance of a tzinfo subclass. The remaining arguments may be ints.
# File: C:\Anaconda\lib\datetime.py
# Type: type
```

In [5]: dt.datetime(month = 3, day = 3, second = 59)

 报错信息如下:TypeError: Required argument 'year' (pos 1) not found。原因分析:year 为必选参数。

Out [5]:
```
---------------------------------------------------------------------------
TypeError                                 Traceback (most recent call last)
<ipython-input-5-f04f4089a039> in <module>()
----> 1 dt.datetime(month = 3, day = 3, second = 59)
      2
      3

TypeError: Required argument 'year' (pos 1) not found
```

In [6]: dt.datetime(year = 2018, month = 3, day = 3, second = 59)

 提示 | minute、hour 为可选参数，可以省略。

Out [6]: datetime.datetime(2018, 3, 3, 0, 0, 59)

39.3 转换方法

In [7]:
 提示 | 在人们的日常生活中，表示日期和时间的方式有很多种，如"3th of July,2018""2019-1-3"和"2018-07-03 00：00：00"等。

#前两种属于非标准格式，不能作为 dt.datetime()的参数。
#最后一种属于标准格式，可以作为 dt.datetime()的参数。

 思路 | 将非标准格式的时间数据转换为标准格式的时间数据的方法有两种。

#第一种方法：用 dateutil 包中的方法 parser. parse()
#第二种方法：用 pandas 包中的方法 to_datetime()

In [8]:
#例1
dt.datetime("3th of July,2018")

 提示 | 报错信息：TypeError: an integer is required (got type str)。
原因分析："3th of July ,2018"为非标准格式时间数据，不能作为 dt.datetime()的参数。

Out [8]: ---------------------------
TypeError   Traceback (most recent call last)
<ipython-input-8-3e23f079018c> in <module>()
      1 #例1
----> 2 dt.datetime("3th of July,2018")

TypeError: an integer is required (got type str)

39 日期与时间

In [9]:
```
#例2
dt.datetime("2019-1-3")
```

报错信息：TypeError: an integer is required (got type str)。
原因分析："2018-1-1"为非标准格式时间数据，不能作为 dt.datetime()的参数。

Out [9]:
```
---------------------------------------------------------------------------
TypeError                                 Traceback (most recent call last)
<ipython-input-9-8c4e4a774449> in <module>()
      1 #例2
----> 2 dt.datetime("2019-1-3")
      3

TypeError: an integer is required (got type str)
```

Python中的日期数据的规范表示方法——是否为规范表示

```
√ 2018-07-03 00:00:00
                                    | parser.parse("3th of July, 2018")    √
× 3th of July,2018  — 转换方法→      | pd.to_datetime("3th of July, 2018")  √
                                    | parser.parse("2019-1-3")             √
× 2019-1-3          — 转换方法→      | pd.to_datetime("2019-1-3")           √
```

In [10]:

（1）用 dateutil 包中的 parser()进行日期格式的转换。

```
from dateutil import parser
date = parser.parse("3th of July,2018")
print(date)
```

Out [10]: 2018-07-03　00:00:00

In [11]:
```
date = parser.parse("2019-1-3")
print(date)
```

Out [11]: 2019-01-03　00:00:00

In [12]:

 （2）用 pandas 包中的 to_datetime()方法进行格式转换。

```
import pandas as pd
pd.to_datetime("3th of July, 2018")
```

 注意 pd.to_datetime()的返回值的类型为时间戳（Timestamp）。

Out [12]: Timestamp('2018-07-03 00:00:00')

In [13]:
```
import pandas as pd
pd.to_datetime("2019-1-3")
```

Out [13]: Timestamp('2019-01-03 00:00:00')

39.4 显示系统当前时间

In [14]:

 提示 显示系统当前时间——方法 datetime.now()。

```
dt.datetime.now()
```

Out [14]: datetime.datetime(2018, 5, 24, 21, 39, 50, 155634)

In [15]:

 提示 显示系统当前时间——方法 datetime.today()。

```
dt.datetime.today()
```

Out [15]: datetime.datetime(2018, 5, 24, 21, 39, 50, 913872)

In [16]:

 技巧 显示周几的方法——在 strftime(String Format Time)中进行设置。

39 日期与时间

```
now = dt.datetime.now()
now.strftime("%W"),now.strftime("%a"),now.strftime("%A"),now.strf
time("% B"),now.strftime("%C"),now.strftime("%D")
```

 注意：参数的大小写所代表的含义不同，如%a 和%A 的含义不同。

```
# %a     本地简化星期名称
# %A     本地完整星期名称
# %b     本地简化的月份名称
# %B     本地完整的月份名称
# %c     本地相应的日期表示和时间表示
# %W     一年中的星期数（00~53）
```

Out [16]: ('21', 'Thu', 'Thursday', 'May', '20', '05/24/18')

39.5 计算时差

In [17]:

 提示：计算时差的方法——用减法。

```
d1 = dt.datetime.now()
d2 = dt.datetime(year = 2017, month = 3, day = 3)
(d1-d2).days
```

 提示：在此，.days 属性为计算单位。

Out [17]: 447

39.6 时间索引

In [18]:

 提示：Pandas 中的时间可以作为索引，具体方法——用 Pandas 中的方法 DatetimeIndex()。

```python
Index = pd.DatetimeIndex(["2018-1-1","2019-1-2","2018-1-3","2018-1-4","201 8-1-5"])
data = pd.Series([1,2,3,4,5],index = Index)
data
```

Out [18]:
```
2018-01-01    1
2019-01-02    2
2018-01-03    3
2018-01-04    4
2018-01-05    5
dtype: int64
```

In [19]:
```python
data["2018-1-2"]
```

Out [19]: Series([], dtype: int64)

In [20]:
```python
data["2018"]    #此时已过滤掉了 2019 年的数据
```

Out [20]:
```
2018-01-01    1
2018-01-03    3
2018-01-04    4
2018-01-05    5
dtype: int64
```

In [21]:
```python
data-data["2018-1-4"]
```

Out [21]:
```
2018-01-01    NaN
2018-01-03    NaN
2018-01-04    0.0
2018-01-05    NaN
2019-01-02    NaN
dtype: float64
```

In [22]:
```python
data
```

Out [22]:
```
2018-01-01    1
2019-01-02    2
2018-01-03    3
```

39 日期与时间

```
2018-01-04    4
2018-01-05    5
dtype: int64
```

In [23]: `data.to_period(freq = "D")`

 提示 在此,"freq = "D""代表的是时间单位为 Day(天)。

Out [23]:
```
2018-01-01    1
2019-01-02    2
2018-01-03    3
2018-01-04    4
2018-01-05    5
Freq: D, dtype: int64
```

In [24]: `data.to_period(freq = "M")`

 提示 在此,"freq = "M""代表的是时间单位为 Month(月)。

Out [24]:
```
2018-01    1
2019-01    2
2018-01    3
2018-01    4
2018-01    5
Freq: M, dtype: int64
```

In [25]: `data−data[3]`

 注意 2019 年数据的计算结果如下。

Out [25]:
```
2018-01-01    −3
2019-01-02    −2
2018-01-03    −1
2018-01-04     0
2018-01-05     1
```

In [26]:
```
data-data["20180104"]
```

 以上计算中,日期作为显式 index,是计算的索引依据,并不参加计算,参加计算的是这些索引代表的 value。

Out [26]:
```
2018-01-01    NaN
2018-01-03    NaN
2018-01-04    0.0
2018-01-05    NaN
2019-01-02    NaN
dtype: float64
```

39.7 period_range()函数

In [27]:

 与 Python 基础语法中的 range()函数和 NumPy 中的 arange()函数类似,Pandas 中有 period_range()函数。

```
pd.period_range("2019-1", periods = 10, freq = "D")
```

 period_range()函数的主要参数如下。

```
#freq:代表的是时间单位,Y、M、D分别代表的是年、月、日
#periods: 代表的是时间单位的个数
#第一个参数为起始时间
```

Out [27]: PeriodIndex(['2019-01-01', '2019-01-02', '2019-01-03','2019-01-04',
                     '2019-01-05', '2019-01-06','2019-01-07','2019-01-08',
                     '2019-01-09', '2019-01-10'],dtype = 'period[D]', freq = 'D')

In [28]:
```
pd.period_range("2019-1", periods = 10, freq = "M")
```

Out [28]: PeriodIndex(['2019-01', '2019-02', '2019-03', '2019-04', '2019-05',
                     '2019-06','2019-07','2019-08','2019-09','2019-10'],
                     Dtype = 'period[M]', freq = 'M')

40 可视化

Q&A 常见疑问及解答。

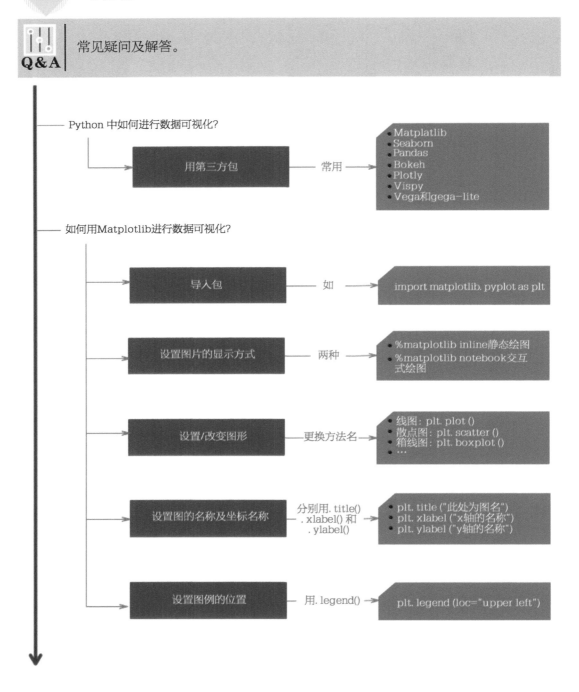

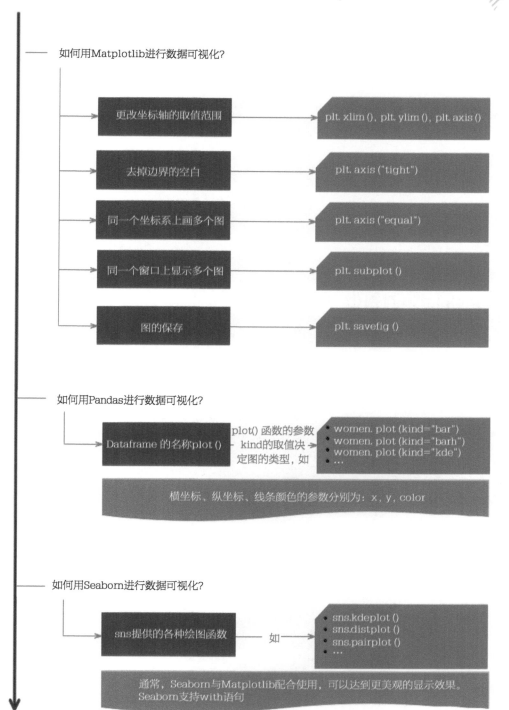

40 可视化

In [1]:

 数据分析和数据科学项目中常用于数据可视化的 Python 包如下。

```
#Matplotlib
#Seaborn
#Pandas
#Bokeh
#Plotly
#Vispy
#Vega
#gega-lite
```

40.1 Matplotlib 可视化

In [2]:

 Matplotlib 提供了 Matlab 风格的绘图工具,常用模块有两种。

```
#(1)绘图API：pyplot,在Jupyter Notebook中通常用于可视化,本书
   采用的是matplotlib.pyplot
#(2)集成库：pylab,是Matplotlib和SciPy、NumPy的集成库

import matplotlib.pyplot as plt
```

 在此导入的是 matplotlib.pyplot,而不是 Matplotlib。

```
%matplotlib inline
```

 Matplotlib 的画图方式可分为两种：inline 和 notebook。

```
#(1)inline为静态绘图,嵌入Jupyter Notebook中显示
#(2)notebook为交互式图,在Jupyter Notebook中显示的图可以支持
    一定的用户交互,如关闭图片显示、缩放图片等。
```

In [3]:
```
import pandas as pd
women = pd.read_csv('women.csv')
women.head()
```

 注意 从输出结果可以看出，women 中多了一个新列 "Unnamed: 0"，也就是说，数据在读入过程中发生了变化。

Out [3]:

	Unnamed: 0	height	weight
0	1	58	115
1	2	59	117
2	3	60	120
3	4	61	123
4	5	62	126

In [4]:

 提示 原因分析：pd.read_csv()在读入数据时自动增加了一个 index 列，即参数 index_col 的默认值为 1。
纠正方法：将 index_col 设为 0。

```
women = pd.read_csv('women.csv', index_col = 0)
women.head()
```

Out [4]:

	height	weight
1	58	115
2	59	117
3	60	120
4	61	123
5	62	126

In [5]:

 提示 在二维坐标系上绘图：plt.plot()。

```
plt.plot(women["height"], women["weight"])
plt.show()
```

 注意 如果不写 plt.show()，则显示
[<matplotlib.lines.Line2D at 0x2064770b550>]

40 可视化

Out [5]:

In [6]:

> 💡 **提示** 生成一个试验数据集 t，用于后续操作中的可视化处理。

```
import numpy as np
t = np.arange(0., 4., 0.1)
t
```

Out [6]: array([0. , 0.1, 0.2, 0.3, 0.4, 0.5, 0.6, 0.7, 0.8, 0.9, 1. , 1.1, 1.2, 1.3, 1.4, 1.5, 1.6, 1.7, 1.8, 1.9, 2. , 2.1, 2.2, 2.3, 2.4, 2.5, 2.6, 2.7, 2.8, 2.9, 3. , 3.1, 3.2, 3.3, 3.4, 3.5, 3.6, 3.7, 3.8, 3.9])

In [7]:

> 💡 **提示** 图中显示多个线条的方法为：在 plt.plot() 中写多个参数，参数格式为 "x,y1,x,y2,x,y3,x,y4,…"。

```
plt.plot(t,t,t,t+2,t,t**2,t,t+8)
```

> 💡 **提示** 参数 "t,t,t,t+2,t,t**2,t,t+8" 的含义为 "(t,t),(t,t+2),(t,t**2),(t,t+8)"，每个括号中的数值分别为 X 轴和 Y 轴，说明同一个 x 对应多个 y。

```
plt.show()
```

Out [7]:

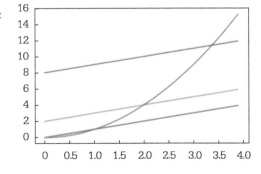

40.2 改变图的属性

In [8]:

> 提示 （1）设置"点"的类型——在 plt.plot()中增加第三个实参的取值，如"o"。

```
plt.plot(women["height"], women["weight"],"o")
plt.show()
```

Out [8]:

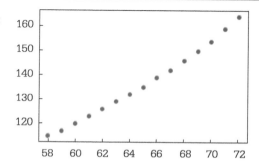

In [9]:

> 提示 （2）设置线的颜色与形状——修改 plt.plot()的第三个实参。

```
plt.plot(women["height"], women["weight"],"g--")
plt.show()
```

Out [9]:

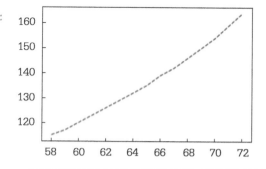

In [10]:
```
plt.plot(women["height"], women["weight"],"rD")
```
> 提示 "rD" 的含义为"红色（red）+钻石（diamond）"。更多实参，请参考 Matplotlib 的官网说明文档。
```
plt.show()
```

可以通过帮助文档进一步了解 plt.plot()的第三个参数的含义,具体命令为:plt.plot?

Out [10]:

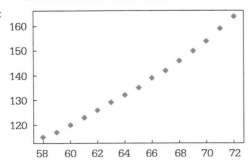

In [11]:

 (3)显示汉字的方法:用 plt.rcParams()设置汉字字体。

 应在调用 plt.plot()之前设置字体,否则汉字显示出现乱码现象。

plt.rcParams['font.family'] = "SimHei"

 在 Maplotlib 中,汉字显示经常遇到乱码现象,原因是没有设置汉字字体或所设置的汉字字体在本机上未找到。查看本机上的字体集的方法为:
import matplotlib.font_manager
[f.name for f in matplotlib.font_manager.fontManager.ttflist]
此外,如果坐标系中负号的显示为乱码,那么再增加一行代码:
plt.rcParams['axes.unicode_minus']=False

plt.plot(women["height"], women["weight"],"g--")
plt.title("此处为图名")
plt.xlabel("X轴的名称")
plt.ylabel("Y轴的名称")

plt.show()

提示 | plt.title()、plt.xlabel()和 plt.ylabel()分别对应的是图的标题、X轴名称和Y轴名称。

Out [11]:

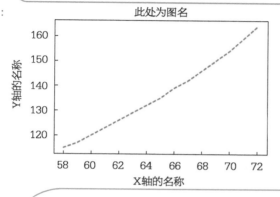

In [12]:

 提示 | （4）设置图名以及X/Y轴名称的修改，分别采用三个方法/函数：plt.title()、plt.xlabel()和 plt.ylabel()。

```
plt.rcParams['font.family'] = "SimHei"  #显示汉字的方法
plt.plot(women["height"], women["weight"],"g--")
plt.title("此处为图名")
plt.xlabel("X轴的名称")
plt.ylabel("Y轴的名称")

plt.show()
```

 注意 | plt.title()、plt.xlabel()和 plt.ylabel()的位置必须在 plot()和 show()之间。

Out [12]:

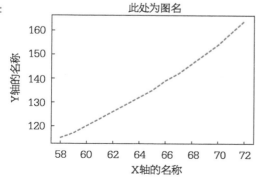

40 可视化

In [13]:

（5）图例的位置：plt.legend(loc = "位置")。

```
plt.rcParams['font.family'] = "SimHei"

plt.plot(women["height"], women["weight"],"g--")
plt.title("此处为图名")
plt.xlabel("X轴的名称")
plt.ylabel("Y轴的名称")

plt.legend(loc = "upper left")
```

loc = "upper left"的含义为"图例放在左上角"。关于 loc 参数的更多值，请参见 plt.legend 的帮助信息。

```
plt.show()
```

Out [13]:

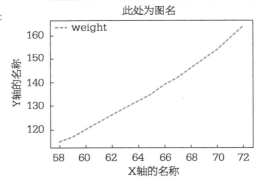

40.3 改变图的类型

In [14]:

将 plt.plot()函数替换为其他函数，如 plt.scatter()。更多函数请参见 Matplotlib 的官网。

建议读者访问 Matplotlib 官网的示例栏目（https://matplotlib.org/gallery/index.html），可以查看其各种显示效果，可以查看每个示例图的源代码。

```
plt.scatter(women["height"], women["weight"])
```

 plt.scatter()的功能为绘制散点图。

```
plt.show()
```

Out [14]:

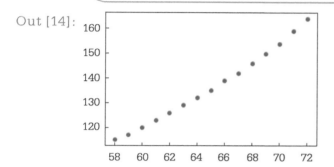

In [15]:
```
%matplotlib inline
plt.scatter(women.height, women.weight)
```

Out [15]: <matplotlib.collections.PathCollection at 0x21b93e51400>

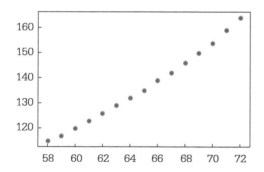

40.4 改变图的坐标轴的取值范围

In [16]: 横坐标和纵坐标的定义：分别采用 plt.xlim()和 plt.ylim()。
横坐标和纵坐标同时定义：plt.axis()。

40 可视化

```
import matplotlib.pyplot as plt
import numpy as np
%matplotlib inline
```

 提示 先生成试验数据 x，用于后续可视化处理。

```
x = np.linspace(0,10,100)
```

 提示 "np.linspace(0,10,100)" 的功能是返回一个含有 100 个元素且每个元素的取值范围为[0,10]的等距离数列，见本书【35 随机数】。

```
plt.plot(x,np.sin(x))

plt.xlim(11,-2)
```

 提示 plt.xlim(11,-2)的含义为"X轴的取值范围为[11,-2]"。

```
plt.ylim(2.2,-1.3)
```

 提示 plt.ylim(2.2,-1.3)的含义为"Y轴的取值范围为[2.2,-1.3]"。

Out [16]:　(2.2, -1.3)

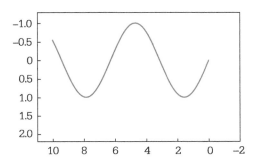

In [17]:
```
plt.plot(x,np.sin(x))
plt.axis([-1,21,-1.6,1.6])
```

 横坐标和纵坐标同时定义:plt.axis(a1, a2, b1, b2)。其中,a1 和 a2 为 X 轴的取值范围,b1 和 b2 为 Y 轴的取值范围。

Out [17]:   [-1, 21, -1.6, 1.6]

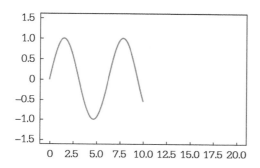

In [18]:
```
plt.plot(x,np.sin(x))
plt.axis([-1,21,-1.6,1.6])
plt.axis("equal")
```

 plt.axis("equal")的含义为 X 轴和 Y 轴的显示长度比例相等。

Out [18]:   (-0.5, 10.5, -1.0993384025373631, 1.0996461858110391)

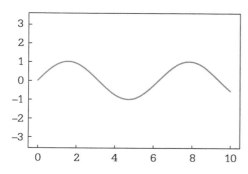

40.5 去掉边界的空白

In [19]:

 提示 | 去掉边界的空白的方法为：
plt.axis("tight")

plt.plot(x,np.sin(x))
plt.axis([-1,21,-1.6,1.6])
plt.axis("tight")

Out [19]: (-0.5, 10.5, -1.0993384025373631, 1.0996461858110391)

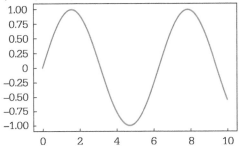

40.6 在同一个坐标上画两个图

In [20]:

 思路 | 分别定义 label（标签），然后用 plt.legend()显示多个 labels。

plt.plot(x,np.sin(x),label = "sin(x)")
plt.plot(x,np.cos(x),label = "cos(x)")
plt.axis("equal")
plt.legend()   #如果没有此行，labels名称将无法显示

Out [20]: <matplotlib.legend.Legend at 0x21b92a6dbe0>

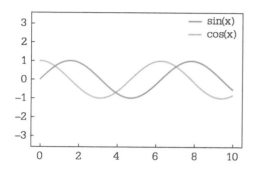

In [21]:
```
plt.plot(x,np.sin(x),label = "sin(x)")
plt.plot(x,np.cos(x),label = "cos(x)")
```

Out [21]: [<matplotlib.lines.Line2D at 0x21b93d6f630>]

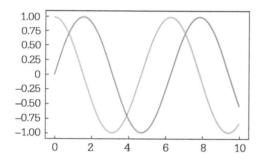

40.7 多图显示

In [22]:

 思路 每个 plot 之前加一个 subplot(x,y,z)。

```
#x：行数
#y：列数
#z：接下来的 plot()的画图位置，即放在第几个"子窗口"中
plt.subplot(2,3,5)
```

 注意 plt.subplot()中，窗口编号是从 1 开始的，并采取"以行优先"的编号方式。另，建议调用 plt.subplot()前采用 plt.figure()设置图片对象的大小（如 plt.figure(figsize=[10,10]），图片大小需要适中，否则显示效果差。

 提示 plt.subplot(2,3,5)的含义为"接下来的显示位置是 2×3 个窗口的第 5 个子窗口"。

40 可视化

```
plt.scatter(women["height"], women["weight"])
plt.subplot(2,3,1)
```

 提示 ｜ plt.subplot(2,3,1)的含义为"接下来的显示位置是2×3个窗口的第1个子窗口"。

```
plt.scatter(women["height"], women["weight"])
plt.show()
```

Out [22]: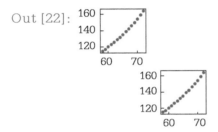

40.8 图的保存

In [23]:

 思路 ｜ 用 Matplotlib 提供的函数，如 plt.savefig()。

```
women = pd.read_csv('women.csv')
plt.plot(women.height, women.weight)
plt.show()
plt.savefig("sagefig.png")
```

 提示 ｜ 以"sagefig.png"为文件名，保存在"当前工作目录"中。

Out [23]: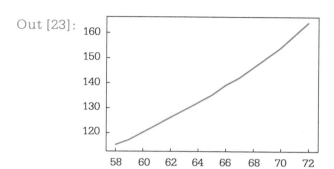

40.9 散点图的画法

In [24]:

> 提示　首先，生成即将用于可视化的试验数据集 X 和 y。
> 生成方法：Python 的 sklearn.datasets.samples_generator 模块提供的函数 make_blobs()。

```
from sklearn.datasets.samples_generator import make_blobs
X,y = make_blobs(n_samples = 300, centers = 4, random_state = 0,
cluster_std = 1.0)
```

> 提示　make_blobs：生成符合正态分布的随机数据集。

```
#参数
    #n_samples:样本数量，即行数
    #n_features:每个样本的特征数量，即列数
    #centers: 类别数
    #random_state: 随机数的生成方式
    #cluster_std: 每个类别的方差
#返回值，有两个:
    #X: 特征集，类型为数组，形状为[n_samples, n_features]
    #y: 每个成员的标签（label），也是个数组，形状为[n_samples]的数组
plt.scatter(X[:,0],X[:,1],c = y,s = 50,cmap = "rainbow")
```

40 可视化

 提示 plt.scatter()函数的参数如下。

#X[:,0]和X[:,1]分别为x坐标和y坐标
#c为颜色
#s为点的大小
#cmap为色带，是c的补充

 提示 X[:,0]的含义为读取数据框 X 的第 0 列，参见本书【38 DataFrame（P274）】。

Out [24]: <matplotlib.collections.PathCollection at 0x21b9484ee80>

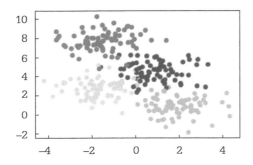

40.10 Pandas 可视化

In [25]:

 思路 在数据分析和数据科学项目中，还可以直接用 Pandas 的画图函数，它继承和优化了 Matplotlib，使 DataFrame（数据框）类数据的可视化处理更容易。

```
import pandas as pd
women = pd.read_csv('women.csv', index_col = 0)
```

 提示 上一行代码的含义：用 Pandas 的方法 read_csv()，将外部文件 women.csv 读入本地数据库 women。

```
women.plot(kind = "bar")
```

>
> **思路**
> Pandas中可以对数据框进行可视化，具体方法：数据框名.plot()。其参数 kind 决定图的类型。kind 的取值见帮助文档：women.plot?

Out [25]:

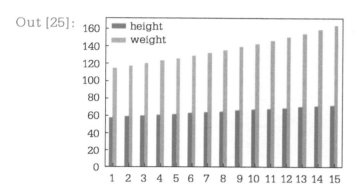

In [26]: ```python
women.plot(kind = "barh")   #barh代表的是横向柱状图
```

Out [26]:

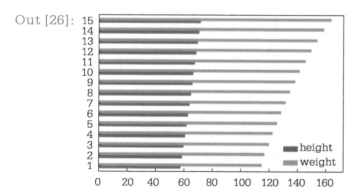

In [27]: ```python
women.plot(kind = "bar", x = "height", y = "weight", color = "g")
```

40 可视化

Out [27]:

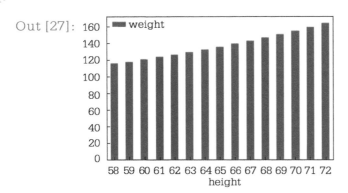

In [28]:
```
women.plot(kind = "kde")   #kde代表的是核密度估计曲线
```

Out [28]:

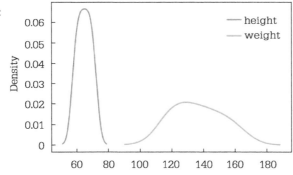

In [29]:
```
women.plot(kind = "bar", x = "height", y = "weight", color = "g")
plt.legend(loc = "best")   #图例位置为"最优"
```

Out [29]:

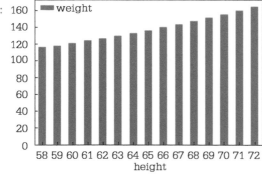

40.11 Seaborn 可视化

In [30]:
```
import pandas as pd
import seaborn as sns
sns.set(style="ticks")
df_women = pd.read_csv('women.csv', index_col=0,header=0)

sns.lmplot(x="height", y="weight", data=df_women)
```

Out [30]: <seaborn.axisgrid.FacetGrid at 0x2ad7c1526c8>

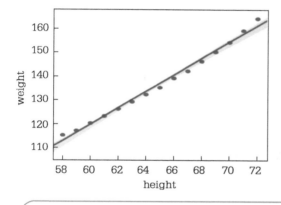

 在 Seaborn 中绘制 plot 图的函数名为 lmplot，与 Matplotlib 中的函数名不同，且调用方式也有所不同。

In [31]:
```
sns.kdeplot(women.height, shade = True)
```

 核密度估计图（Kernel Density Estimation, KDE）。

Out [31]: <matplotlib.axes._subplots.AxesSubplot at 0x21b94b3fc18>

40 可视化

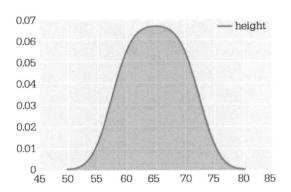

In [32]:

```
sns.distplot(women.height)
```

> 提示：sns.distplot()的功能为绘制displot图。displot图的功能为"直方图+ kdeplot"。

Out[32]: <matplotlib.axes._subplots.AxesSubplot at 0x21b94a14710>

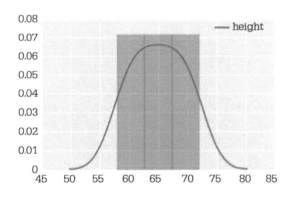

In [33]:

```
sns.pairplot(women)
```

> 提示：sns.pairplot()的功能为绘制散点图矩阵。

Out[33]: <seaborn.axisgrid.PairGrid at 0x21b948b4a20>

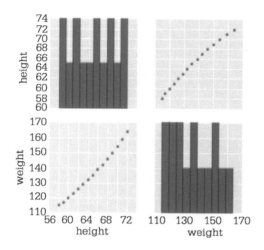

In [34]:
```
sns.jointplot(women.height, women.weight, kind = "reg")
```

 提示　函数 sns.jointplot() 的功能为绘制"联合分布图"。

Out [34]:   <seaborn.axisgrid.JointGrid at 0x21b94f5d7f0>

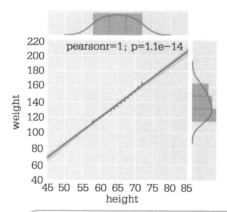

In [35]:

 技巧　可以加一个 with 语句，进行异常处理和句柄释放等功能。

 注意　with 后面有一个冒号。

40 可视化

```
with sns.axes_style("white"):
    sns.jointplot(women.height,women.weight,kind = "reg")
```

Out [35]:

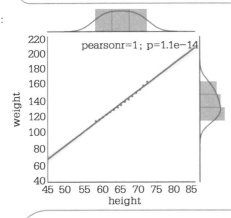

In [36]:

 技巧 Seaborn 还可以放在 for 语句中,将多个变量画在一起。

```
for x in["height","weight"]:
    sns.barplot(x=women[x],y=women[x], palette="vlag")
```

 提示 sns.barplot()的功能为绘制条形图,更多内容见 Seaborn 官网中的 barplot 图示例:https://seaborn.pydata.org/examples/color_palettes.html。

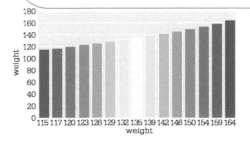

40.12 数据可视化实战

(1) 数据准备

In [37]:

 思路 查看当前工作目录,并将数据文件 salaries.csv 放在当前工作目录中。

```
import os
os.getcwd()
```

提示 读者可以在本书配套资源中找到数据文件 salaries.csv。

注意 读者的"当前工作目录"不一定与本书一样,请以自己计算机上的 Out[40]中的显示结果为准。

Out[37]:   'C:\\Users\\soloman\\clm'

In [38]:

思路 用 Pandas 的 read_csv()函数将文件 salaries.csv 读入内存的对象 salaries 中。

```
import pandas as pd
salaries = pd.read_csv('salaries.csv', index_col = 0)
```

提示 index_col = 0 的含义为,在准备读入的数据文件 salaries.csv 中带有索引列且索引列位于第 0 列。

In [39]:

思路 查看内存对象 salaries 的前 6 行,该对象属于 Pandas 的 DataFrame 类型,参见本书【38 DataFrame(P274)】。

```
salaries.head(6)
```

Out[39]:

	rank	discipline	yrs.since.phd	yrs.service	sex	salary
1	Prof	B	19	18	Male	139750
2	Prof	B	20	16	Male	173200
3	AsstProf	B	4	3	Male	79750
4	Prof	B	45	39	Male	115000
5	Prof	B	40	41	Male	141500
6	AssocProf	B	6	6	Male	97000

40 可视化

（2）导入 Python 包

In [40]:

 为了数据可视化，在此导入 Seaborn 模块，并取别名为 sns，参见本书【26 模块（P170/In[3]）】。

```
import seaborn as sns
```

（3）可视化绘图

In [41]:

 设置 Seanborn 的绘图样式或主题为 darkgrid（灰色+网格）。

```
sns.set_style('darkgrid')
```

 用 sns.stripplot() 绘制散点图。

```
sns.stripplot(data = salaries, x = 'rank', y = 'salary', jitter = True, alpha = 0.5)
```

 data 为数据来源；x 和 y 分别用于设置 X 轴和 Y 轴；jitter 的含义是散点是否有抖动（注：增加随机噪声）；alpha 为透明度。

 继续绘制箱线图。

```
sns.boxplot(data = salaries, x = 'rank', y = 'salary')
```

 data 为数据来源；x 和 y 分别用于设置 X 轴和 Y 轴。

Out[41]: <matplotlib.axes._subplots.AxesSubplot at 0x288151c66d8>

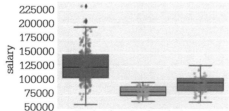

41 Web 爬取

常见疑问及解答。

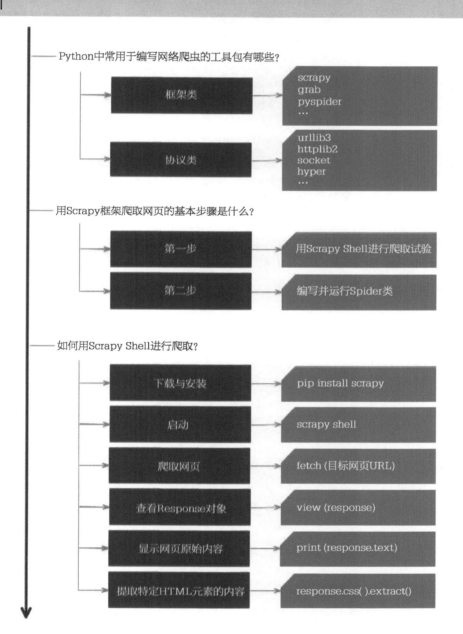

41 Web 爬取

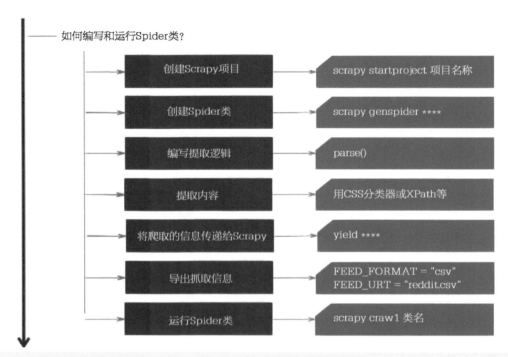

 Python 中爬取 Web 信息非常容易，可借助的第三方包或模块也很多。本书以 Scrapy 为例，重点介绍 Web 爬取的基本流程和关键技术。Scrapy 的基本流程如图 41-1 所示。

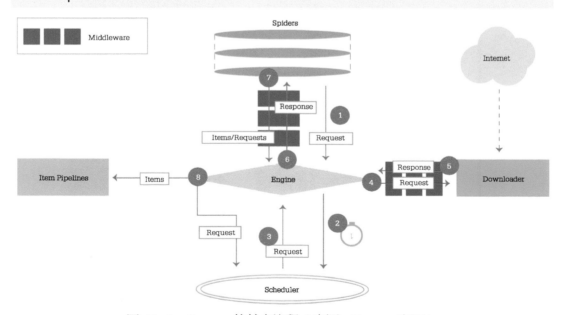

图 41-1　Scrapy 的基本流程（来源：Scrapy 官网）

Scrapy 是基于 HTTP 的 Request（请求）/Response（响应）模式运行的，其数据流由执行引擎（Engine）控制，具体步骤如下。

[1] 执行引擎从爬虫（Spider）处获得一个需要爬取的初始 Request。
[2] 执行引擎调度在调度器（Sheduler）中的 Request 并确定下一个需要爬取的 Request。
[3] 调度器将下一个爬取的 Request 提交给执行引擎。
[4] 执行引擎通过下载器中间件（Downloader Mildware）将 Request 发送至下载器（Downloader）。
[5] 下载结束之后，下载器将下载结果以 Response 对象形式返回给执行引擎（Engine）。
[6] 执行引擎接收 Response 对象并将其发送至爬虫进行处理。
[7] 爬虫负责解析 Response 对象，并将已删除的 Items 和需要跟踪的新 Request 返回到引擎。
[8] 引擎将已处理的 Items 发送到 Items 管道（Items Pipeline），同时将已处理的 Request 报告给调度器，并询问是否还有继续爬取的 Request。
[9] 重复上述步骤，直到调度程序不再有任何未处理的请求为止。

 Scrapy是一种可用于大规模网络数据爬取的Python框架，不仅支持高效爬取Web信息，还提供了将爬取数据处理成结构化数据的工具集。

 Techcrunch是一个科技类博客网站，如图41-2所示。该网站提供了自己的RSS源码：https://techcrunch.com/feed/，如图41-3所示。本章爬取内容为该网站博文的文章标题、文章链接、发布日期和作者姓名等四项内容。

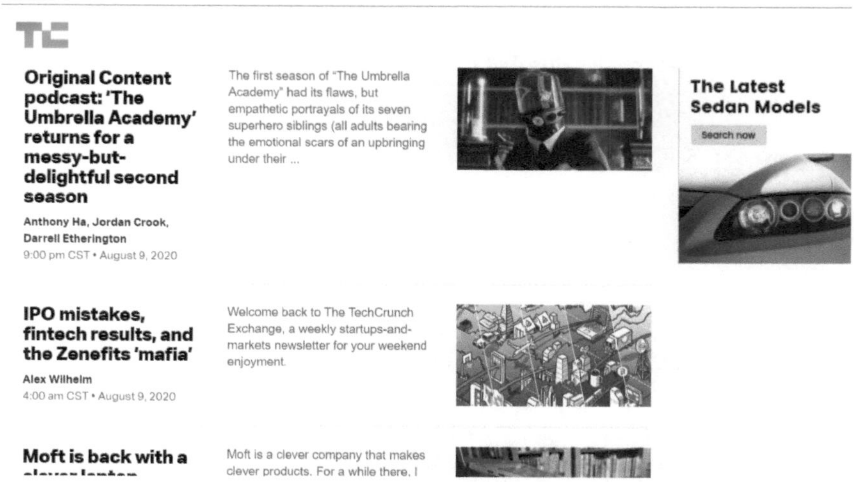

图 41-2　Techcrunch 官网

41 Web 爬取

图 41-3 Techcrunch 官网 RSS 源码

从图41-3可以看出，该网站中博客内容的标记特点为：

- 每篇文章的信息显示在<item></item>标签中
- 文章标题记录在<title></title>标签中
- 文章链接记录在<link>标签中
- 发布日期记录在<pubDate>中
- 作者姓名记录在<dc:creator>中

41.1 Scrapy 的下载与安装

可以用 Conda 工具，参见本书【27 包（P175/In[2]）】。

conda install scrapy 或 conda install -c conda-forge scrappy

与其他章节不同的是，本章代码的输入位置并非为 Jupyter Notebook，而是 Anaconda Prompt。

 也可以使用 PIP 工具进行下载和安装。

pip install scrapy

41.2 Scrapy Shell 的基本原理

 Scrapy Shell 是一种交互式终端，支持在未启动 Spider 的情况下尝试及调试爬取代码。在编写 Spider 时，该终端提供了交互式测试代码的功能，可以避免每次修改后都需要重新运行 Spider 的麻烦。

 Scrapy Shell 的启动方法：在 Anaconda Prompt 中输入如下命令。

scrapy shell

 接着，用 Scrapy Shell 爬取 Techcrunch 网页。

fetch('https://techcrunch.com/feed/')

 当使用 Scrapy Shell 进行爬取时，会返回一个 Response 对象，其中包含了下载的信息，查看下载内容的方法如下。

view(response)

41 Web 爬取

> **提示** 同时，将打开已下载的网页。

41.3 Scrapy Shell 的应用

> **提示** 根据网页标记语言的不同，可以选择 CSS 或 Xpath 等方法解析和提取所需内容。考虑到本章下载内容为 XML 文件，可以采用 XPath 技术对网页信息进行解析和提取。以文章标题信息（<title>元素）的提取为例：

response.xpath("//item/title").extract ()

上一行代码输出了所有爬取的标题。如果仅提取第一个标题,那么命令如下。

```
response.xpath("//item/title/text()").extract_first()
```

 此处,text() 的含义为仅提取元素的 text 属性。
在 XPath 中,"//item/title/text()"的含义为查找 item 元素,从 title 子元素中提取 text 部分。

 同理,对于文章链接以及发布日期的 XPath 提取:

- Link–//item/link/text()
- Date of publishing–//item/pubDate/text()

 一个特殊情况,提取作者信息应为 response.xpath("//item/creator/text()").extract_first(),但是 Scrapy Shell 并没有显示结果(显示结果为空)。

 为此,需要进一步查看和分析<creator>标签:

41 Web 爬取

 可见，<creator>标签含有 "dc:"，因此使用 XPath 不能提取，而且姓名部分含有 "![CDATA[]]" 相关文本。这些是 XML 的命名空间，需要先删除这些命名空间再提取内容。

response.selector.remove_namespaces()

 再次提取作者姓名信息。

response.xpath("//item/creator/text()").extract_first()

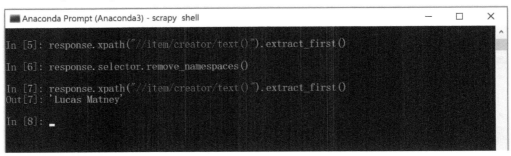

41.4 自定义 Spider 类

 Spider 类是用于爬取网站内容的程序。由于网站设计的多样性，在大规模爬取数据时，需要为不同的网站编写自定义的 Spider 类，还需要编写代码，将爬取的内容转化为结构化格式，如 CSV、JSON、Excel 等，可以在 Scrapy 的内置功能中找到。主要步骤如下：

（1）创建 Scrapy 项目

 注意 创建 Scrapy 项目的前提：调用函数 quit() 退出 Scrapy Shell，再用以下命令创建一个新的 Scrapy 项目。

```
scrapy startproject myProject
```

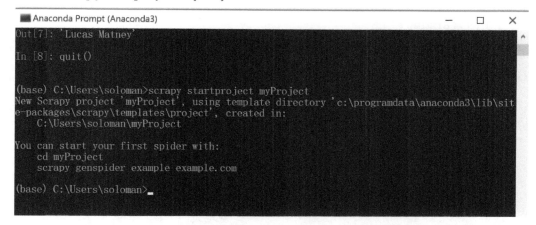

 注意 学会查看系统提示。如先 cd 到 myProject 目录，再输入 scrapy genspider 的命令。

 提示 系统将自动创建 myProject 文件夹，可以通过以下命令查看文件夹结构。

```
tree myProject   /F
```

注意 | 此处，最重要的文件和文件夹如下。

- settings.py 文件——包含项目的设置信息。
- spiders 文件夹——存储所有的自定义 Spider 类。在接下来的操作中，每次运行一个 Spider 类，Scrapy 会自动在此文件夹中查找并运行。

（2）创建 Spider 类

提示 | 接着，进入 myProject 文件夹，并创建一个 Spider 类 techcrunch，主要代码如下。

cd myProject
scrapy genspider techcrunch techcrunch.com/feed/

提示 | 从上图可以看出，在 spiders 文件夹中创建了 techcrunch.py 文件，其内部包含了一个基本模板。

提示 | 该模板的主要内容包括如下。

- name：Spider 类的名字

- allowed_domains：允许此 Spider 类爬取的网页 URL，不属于此域名的 URL 无法爬取
- parse(self, response)：当成功爬取到一个 URL 时，会调用此函数，而且这里的 response 对象和前面提到的一致

 每次爬取成功后，都会调用 parse()函数，该函数是编写提取逻辑的地方。现在添加之前试验过的提取逻辑。

```
def parse(self, response):
    # 删除 XML 命名空间
    response.selector.remove_namespaces()
    # 提取文章信息
    titles = response.xpath('//item/title/text()').extract()
    authors = response.xpath('//item/creator/text()').extract()
    dates = response.xpath('//item/pubDate/text()').extract()
    links = response.xpath('//item/link/text()').extract()
    for item in zip(titles, authors, dates, links):
        scraped_info = {
            'title': item[0],
            'author': item[1],
            'publish_date': item[2],
            'link': item[3]
        }
        yield scraped_info
```

 编写以上代码是注意代码的对齐方式，代码对齐方式如下图所示。

41 Web 爬取

注意 此处，yield scraped_info 将爬取的信息传递给 Scrapy，后者将依次处理信息并存储。

提示 保存 techcrunch.py 文件，打开 CMD 窗口，进入 myProject 文件夹，运行这个 Spider 类。

scrapy crawl techcrunch

```
(base) C:\Users\soloman\myProject>scrapy crawl techcrunch
2020-08-10 08:13:59 [scrapy.utils.log] INFO: Scrapy 1.8.0 started (bot: myProject)
2020-08-10 08:13:59 [scrapy.utils.log] INFO: Versions: lxml 4.4.1.0, libxml2 2.9.9, cssselect 1.1.0, parsel 1.5.2, w3lib 1.21.0, Twisted 19.10.0, Python 3.7.4 (default, Aug  9 2019, 18:34:13) [MSC v.1915 64 bit (AMD64)], pyOpenSSL 19.0.0 (OpenSSL 1.1.1d  10 Sep 2019), cryptography 2.7, Platform Windows-10-10.0.18362-SP0
2020-08-10 08:13:59 [scrapy.crawler] INFO: Overridden settings: {'BOT_NAME': 'myProject', 'NEWSPIDER_MODULE': 'myProject.spiders', 'ROBOTSTXT_OBEY': True, 'SPIDER_MODULES': ['myProject.spiders']}
2020-08-10 08:13:59 [scrapy.extensions.telnet] INFO: Telnet Password: b4b33ae7d1943566
2020-08-10 08:13:59 [scrapy.middleware] INFO: Enabled extensions:
['scrapy.extensions.corestats.CoreStats',
 'scrapy.extensions.telnet.TelnetConsole',
 'scrapy.extensions.logstats.LogStats']
2020-08-10 08:13:59 [scrapy.middleware] INFO: Enabled downloader middlewares:
['scrapy.downloadermiddlewares.robotstxt.RobotsTxtMiddleware',
 'scrapy.downloadermiddlewares.httpauth.HttpAuthMiddleware',
 'scrapy.downloadermiddlewares.downloadtimeout.DownloadTimeoutMiddleware',
 'scrapy.downloadermiddlewares.defaultheaders.DefaultHeadersMiddleware',
 'scrapy.downloadermiddlewares.useragent.UserAgentMiddleware',
```

```
 'downloader/request_count': 5,
 'downloader/request_method_count/GET': 5,
 'downloader/response_bytes': 59822,
 'downloader/response_count': 5,
 'downloader/response_status_count/200': 2,
 'downloader/response_status_count/301': 3,
 'elapsed_time_seconds': 3.111951,
 'finish_reason': 'finished',
 'finish_time': datetime.datetime(2020, 8, 10, 0, 14, 2, 895179),
 'item_scraped_count': 20,
 'log_count/DEBUG': 25,
 'log_count/INFO': 10,
 'response_received_count': 2,
 'robotstxt/request_count': 1,
 'robotstxt/response_count': 1,
 'robotstxt/response_status_count/200': 1,
 'scheduler/dequeued': 3,
 'scheduler/dequeued/memory': 3,
 'scheduler/enqueued': 3,
 'scheduler/enqueued/memory': 3,
 'start_time': datetime.datetime(2020, 8, 10, 0, 13, 59, 783228)}
2020-08-10 08:14:02 [scrapy.core.engine] INFO: Spider closed (finished)

(base) C:\Users\soloman\myProject>
```

以上输出信息显示所有数据已经下载,并且存放在字典中。

 如果服务器按 Robots 协议拒绝数据,那么修改过 settins.py 文件中的 ROBOTSTXT_OBEY 值改为 False 即可。

(3)将爬取的数据保存为 Csv 格式

 Scrapy 可以将爬取的内容导出为 CSV、JSON 等格式,打开 settins.py 文件,并且添加以下信息。

```
# 导出为 CSV 文件
FEED_FORMAT = "csv"
FEED_URI = "techcrunch.csv"
```

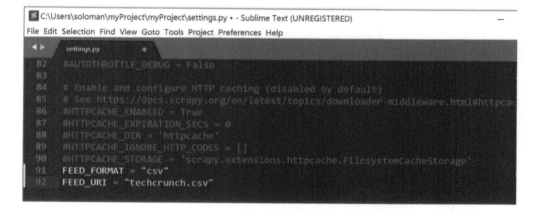

 上图中有如下两个参数。

- FEED_FORMAT:定义输出文件的格式,目前支持的格式有 CSV、JSON、JSON lines、XML。
- FEED_URI:输出文件的路径。

 再次运行这个 Spider 类。

```
scrapy crawl techcrunch
```

41 Web 爬取

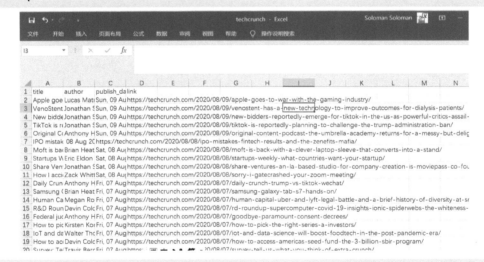

 在 myProject 文件夹中输出 techcrunch.csv 文件,内容如下。

 到目前为止,已经成功地创建了一个系统,可以爬取网页内容,从中提取目标信息,并且保存为结构化格式。

 本章内容并非为作者原创,主要内容修改自 Mohd Sanad Zaki Rizvi 的《Web Scraping in Python Using Scrapy》。

41.5 综合运用

思路 本例综合运用本章学习内容,并将爬取对象从 HTML 网页拓展至 XML 数据。

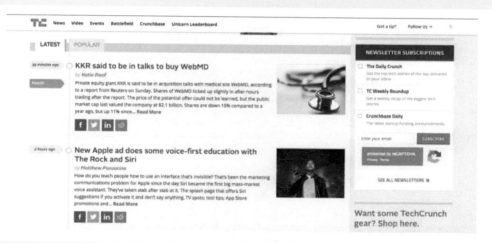

提示 Scrapy 的一大特点就是能够轻松处理 XML 数据,下面从 Techcrunch 的 RSS 源码中爬取信息。查看 Techcrunch 的 RSS 源码,红色框中的部分是需要提取的信息。

41 Web 爬取

提示 从上图中的 XML 代码页面可以看出。

- 每篇文章的信息显示在<item></item>标签中。
- 文章标题保存在<title></title>标签中。
- 文章链接保存在<link>标签中。
- 发布日期保存在<pubDate>标签中。
- 作者姓名保存在<dc:creator>标签中。

（1）Scrapy Shell 调试代码

提示 与本书 41.1~41.4 部分类似，首先在 Scrapy Shell 中调试代码，来爬取网页。

scrapy shell
fetch('https://techcrunch.com/feed/')

提示 在 XML 格式的网页中，提取文章标题、链接及发布日期信息与 response.css() 类似。response.xpath()用于处理 XPath，现在提取文章标题信息。

response.xpath("//item/title").extract_first()

输出结果较多,需进一步提取标题部分。

response.xpath("//item/title/text()").extract_first()

 text()相当于 CSS 采集器中的::text。在 XPath 中,"//item/title/text()"的含义为查找 item 元素,从 title 子元素中提取 text 部分。

 同理,提取链接和发布日期的 XPath。

Link-//item/link/text()
Date of publishing-//item/pubDate/text()

 提取作者姓名信息。

response.xpath("//item/creator/text()").extract_first()

 但是,上一行代码没有返回需要的结果,因此需进一步分析<creator>标签。

 标签含有 "dc:",因此使用 XPath 不能提取,而且姓名部分含有 "![CDATA[]]" 相关文本,是 XML 的命名空间,需要先删除命名空间再提取内容。

response.selector.remove_namespaces()

提示 再次提取作者姓名信息。

```
response.xpath("//item/creator/text()").extract_first()
```

（2）创建 Spider 类

提示 代码调试完成后，从 Scrapy Shell 退出，进入 secondone 文件夹，创建一个新的 Spider 类。

```
scrapy startproject secondone
```

```
cd secondone
scrapy genspider techcrunch techcrunch.com/feed/
```

将调试过的代码添加到 techcrunch.py 文件中,完整的代码如下。

```python
import scrapy
class TechcrunchSpider(scrapy.Spider):
    name = 'techcrunch'
    allowed_domains = ['techcrunch.com/feed/']
    start_urls = ['http://techcrunch.com/feed//']
    # 设置输出 csv 文件的路径
    custom_settings = {
        'FEED_URI': 'tmp/techcrunch.csv'
    }
    def parse(self, response):
        # 删除 XML 命名空间
        response.selector.remove_namespaces()
        # 提取文章信息
        titles = response.xpath('//item/title/text()').extract()
        authors = response.xpath('//item/creator/text()').extract()
        dates = response.xpath('//item/pubDate/text()').extract()
        links = response.xpath('//item/link/text()').extract()
        for item in zip(titles, authors, dates, links):
            scraped_info = {
                'title': item[0],
                'author': item[1],
                'publish_date': item[2],
                'link': item[3]
            }
            yield scraped_info
```

 以上代码中添加了输出 CSV 文件的路径,即 tmp/techcrunch.csv。

 保存 techcrunch.py 文件,打开 CMD 窗口,进入 firstone 文件夹,运行这个 Spider 类。

```
scrapy crawl techcrunch
```

41 Web 爬取

命令提示符窗口输出。

提示 在 tmp/techcrunch.csv 中得到的 CSV 文件的内容如下。

techcrunch.csv 文件内容截图。

提示 本章内容并非作者原创，主要内容修改自 Mohd Sanad Zaki Rizvi 的《Web Scraping in Python Using Scrapy》。

第五篇
数据分析

统计分析
机器学习
自然语言处理
人脸识别与图像分析

42 统计分析

 常见疑问及解答。

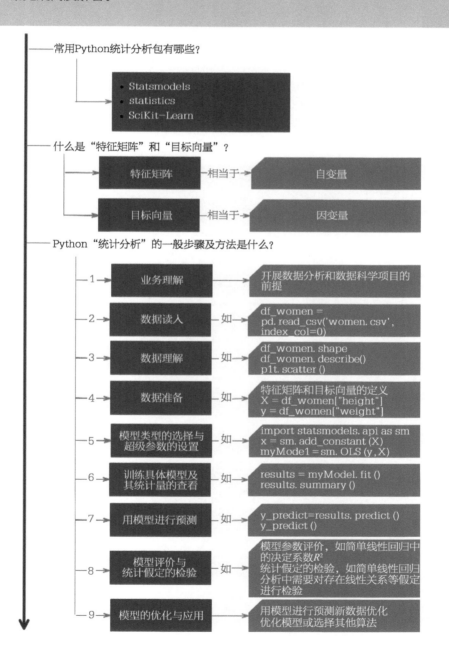

42 统计分析

In [1]:

思路 统计分析中有两个重要概念——特征矩阵和目标向量。下面以 y= F(X)为例。

```
#X:自变量
    #用特征矩阵（Feature_Matrix）表示
    #特征矩阵的行和列分别称为 samples 和 features
    #通常，用 n_samples 和 n_features 分别表示行和列的数量
    #在多数 Python 统计分析模块中，"特征矩阵"应为 NumPy 的 ndarray
     和 Pandas 的 DataFrame 表示，个别模块支持 SciPy 的稀疏矩阵
```

注意 "特征矩阵"中的每个 sample 必须为一行。
"特征矩阵"中不能包含因变量——目标向量。

```
#y:因变量
    #用"目标向量（Target Vector）"表示
```

X为特征矩阵（Feature_Matrix）
→ n个特征（n Features）

↓ m个样本（msamples）

y为目标向量（Target Vector）
→ m个标签（m labels）/样本（m features）

42.1 业务理解

In [2]:

  思路 | "业务理解"是数据分析与数据科学项目的第一步。本例中的主要业务情景如下。

#分析目的：
　　#女性身高与体重之间的关系（能否依据身高预测一位女性的体重？）
#分析数据：
　　#women 的数据集，该数据集源自 The World Almanac and Book of Facts（1975）
　　#给出了年龄在 30~39 岁的 15 名女性的身高和体重数据

42.2 数据读入

In [3]:

  提示 | 建议将准备读入的外部文件放在"当前工作目录"中，在可以读入外部文件之前修改"当前工作目录"。参见本书【34 当前工作目录】。

```
import os
os.chdir(r'C:\Users\soloman\clm')
```

  提示 | 此处，"C:\Users\soloman\clm"为本书作者计算机上的工作路径，建议读者根据自己计算机的实际情况，设置具体工作路径。

```
print(os.getcwd())
```

Out [3]:　　C:\Users\soloman\clm

In [4]:

  提示 | 调用 Pandas 的 read_csv()函数，当读取 CSV 文件时，Pandas 将自动将其转换为一个 DataFrame 对象。

42 统计分析

```
import pandas as pd
import numpy as np
df_women = pd.read_csv('women.csv', index_col=0)
print(df_women.head())
```

 读者可以从本书配套资源中找到数据文件 women.csv。

Out [4]:
```
   height  weight
1      58     115
2      59     117
3      60     120
4      61     123
5      62     126
```

42.3 数据理解

In [5]:

 （1）查看形状，参见本书【38 DataFrame（P279/In[9]）】。

```
df_women.shape
```

Out [5]: (15, 2)

In [6]:

 （2）查看列名，参见本书【38 DataFrame（P278/In[7]）】。

```
print(df_women.columns)
```

Index(['height', 'weight'], dtype = 'object')

In [7]:

 （3）查看描述性统计信息，参见本书【38 DataFrame（P297/In[62]）】。

```
df_women.describe()
```

Out [7]:

	height	weight
count	15.000000	15.000000
mean	65.000000	136.733333
std	4.472136	15.498694
min	58.000000	115.000000
25%	61.500000	124.500000
50%	65.000000	135.000000
75%	68.500000	148.000000
max	72.000000	164.000000

In [8]:

 提示 （4）数据可视化。

```
import matplotlib.pyplot as plt
%matplotlib inline
plt.scatter(df_women["height"], df_women["weight"])
plt.show()
```

Out [8]:

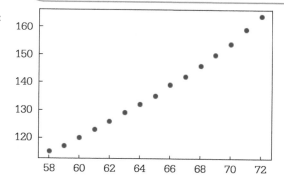

42.4 数据准备

In [9]:

 提示 从可视化结果看出，可以进行线性回归分析。为此，需要准备线性回归分析所需的特征矩阵（X）和目标向量（y）。

```
X = df_women["height"]
y = df_women["weight"]
```

42 统计分析

```
In [10]:   X
Out [10]:  1     58
           2     59
           3     60
           4     61
           5     62
           6     63
           7     64
           8     65
           9     66
           10    67
           11    68
           12    69
           13    70
           14    71
           15    72
           Name: height, dtype: int64
In [11]:   y
Out [11]:  1     115
           2     117
           3     120
           4     123
           5     126
           6     129
           7     132
           8     135
           9     139
           10    142
           11    146
           12    150
           13    154
           14    159
           15    164
           Name: weight, dtype: int64
```

提示 其实此处的对象 y 并非我们需要的目标向量。读者可以用 type(y)方法查看 y 的数据类型。后续操作中没有报错的原因在于 statsmodels 包进行了自动类型转换。但是,如果用其他包(如 SciKit Learn 等)会报错,需要我们对 y 进行 np.ravel 转换,参见本书【43 机器学习（P393/In[32]）】。

42.5 模型类型的选择与超级参数的设置

In [12]:
```
import statsmodels.api as sm
```

> 💡 **提示** Python 统计分析中常用包有 Statsmodels、statistics、SciKit-Learn。

```
C:\Anaconda\lib\site-packages\statsmodels\compat\pandas.py:56:
FutureWarning: The pandas.core.datetools module is deprecated and will be removed
in a future version. Please use the pandas.tseries module instead.
  from pandas.core import datetools
```

In [13]: X
Out [13]:
```
1     58
2     59
3     60
4     61
5     62
6     63
7     64
8     65
9     66
10    67
11    68
12    69
13    70
14    71
15    72
Name: height, dtype: int64
```

In [14]:
> ⚠️ **注意** OLS 的超级参数：默认情况下，OLS 不含截距项（Intercept），可以通过如下转换方式来设置截距项这一超级参数。

```
X_add_const = sm.add_constant(X)
```

> 💡 **提示** 给X新增一列，列名const，每行取值1.0。

42 统计分析

 不要将此行代码写成"X = sm.add_constant(X)",否则只有第一次执行该 Cell 时才能得到正确结果。当多次执行时,每次执行该 Cell 后,X 的值会不断发生改变。

X_add_const

Out [14]:

	const	height
1	1.0	58
2	1.0	59
3	1.0	60
4	1.0	61
5	1.0	62
6	1.0	63
7	1.0	64
8	1.0	65
9	1.0	66
10	1.0	67
11	1.0	68
12	1.0	69
13	1.0	70
14	1.0	71
15	1.0	72

In [15]:
```
myModel = sm.OLS(y, X_add_const)
```

 sm.OLS()函数的前两个形参分别为endog和exog。在包statsmodels的开发者看来,"y和X对于这个模型来分别是内生的(endog)和外生的(exog)",参见 http://www.statsmodels.org/stable/endog_exog.html。

42.6 训练具体模型及查看其统计量

In [16]:
```
results = myModel.fit()
print(results.summary())
```

Out [16]:
OLS Regression Results

```
==============================================================
Dep. Variable:          weight   R-squared:              0.991
Model:                     OLS   Adj. R-squared:         0.990
Method:          Least Squares   F-statistic:            1433.
Date:        Sun, 19 Aug 2018   Prob (F-statistic):   1.09e-14
Time:                 20:03:13   Log-Likelihood:       -26.541
No. Observations:           15   AIC:                    57.08
Df Residuals:               13   BIC:                    58.50
Df Model:                    1
Covariance Type:     nonrobust
==============================================================
                coef    std err       t    P>|t|    [0.025  0.975]
==============================================================
const        -87.5167    5.937   -14.741   0.000  -100.343  -74.691
height         3.4500    0.091    37.855   0.000     3.253    3.647
==============================================================
Omnibus:              2.396   Durbin-Watson:        0.315
Prob(Omnibus):        0.302   Jarque-Bera (JB):     1.660
Skew:                 0.789   Prob(JB):             0.436
Kurtosis:             2.596   Cond. No.              982.
==============================================================

Warnings:
[1] Standard Errors assume that the covariance matrix of the errors is cor-
rectly specified.
C:\Anaconda\lib\site-packages\scipy\stats\stats.py:1394: UserWarning:
kurtosistest only valid for n>=20 ... continuing anyway, n=15"anyway,
n=%i" % int(n))
```

In [17]:

 提示 | 查看"斜率"和"截距项"。

results.params

Out [17]: const   -87.516667
height   3.450000
dtype: float64

42.7 拟合优度评价

In [18]: 评价回归直线的拟合优度——计算 R^2（决定系数）。

```
results.rsquared
```

 R^2 的取值范围为[0,1]，越接近 1，说明"回归直线的拟合优度越好"。

Out [18]: 0.9910098326857505

42.8 建模前提假定的讨论

In [19]: 在基于统计学方法完成数据分析和数据科学任务时，不仅需要进行模型优度的评价，还需要重点分析统计方法的"应用前提假定"是否成立。

 通常，统计分析都是建立在一个或多个"前提假定条件"之上的。以简单线性回归为例，其"前提假定条件"如下。

```
#（1）X 和 y 之间存在线性关系→检验方法为计算 F 统计量
#（2）残差项（的各期）之间不存在的自相关性→检验方法为计算 Durbin-
      Watson 统计量
#（3）残差项为正态分布的随机变量→检验方法为计算 JB 统计量
```

In [20]: 查看 F 统计量的 p 值。

```
results.f_pvalue
```

377

思路 | F统计量的p值——用于检验X与y之间是否存在线性关系。自变量（X）和因变量（y）之间存在线性关系是"线性回归分析"前提假定之一。

方法 | 通常，查看p值是否小于经验值，如0.05。

Out [20]: 1.0909729585997682e-14

In [21]:

提示 | 显示Durbin-Watson统计量。

sm.stats.stattools.durbin_watson(results.resid)

思路 | Durbin-Watson统计量用于检查残差项之间是否存在的自相关性，残差项（的各期）之间相互独立是线性回归分析的另一个前提假设。

Out [21]: 0.31538037486218 51

In [22]:

提示 | 显示JB统计量及其p值。

sm.stats.stattools.jarque_bera(results.resid)

提示 | 此函数的返回值有4个，分别为JB值、JB的p值、峰度和偏度。

思路 | JB统计量用于检验残差项是否为"正态分布"，残差项属于正态分布是线性回归分析的前提假设之一。

Out [22]: (1.6595730644309743,
0.4361423787323869,
0.7893583826332262,

42 统计分析

2.5963042257390008)

In [23]:

 提示 | 用新模型 results 重新进行预测。

```
y_predict = results.predict()
y_predict
```

Out [23]: array([112.58333333, 116.03333333, 119.48333333, 122.93333333,
       126.38333333, 129.83333333, 133.28333333, 136.73333333,
       140.18333333, 143.63333333, 147.08333333, 150.53333333,
       153.98333333, 157.43333333, 160.88333333])

42.9 模型的优化与重新选择

In [24]:

 提示 | 除了 R^2（决定系数）等统计量，可以通过可视化方法更直观地查看回归效果。

```
plt.rcParams['font.family'] = "simHei" #汉字显示
plt.plot(df_women["height"], df_women["weight"],"o")
plt.plot(df_women["height"], y_predict)
plt.title('女性身高与体重的线性回归分析')
plt.xlabel('身高')
plt.ylabel('体重')
```

Out [24]: Text(0,0.5,'体重')

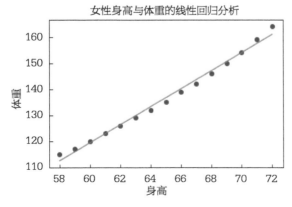

In [25]: 从上图可以看出，本例题用简单线性回归的效果并不好，为此采取多项式回归方法。

In [26]:
```python
import pandas as pd
import numpy as np
df_women = pd.read_csv('women.csv', index_col=0)
X = df_women["height"]
y = df_women["weight"]
X = np.column_stack((X, np.power(X,2), np.power(X,3)))
```

 在多项式分析中，特征矩阵 X 由 3 部分组成，即 X、X 的平方和 X 的三次方。

In [27]:
```python
X_add_const = sm.add_constant(X)
```

 此处，调用 sm.add_constant() 的目的是保留截距项。

```python
X_add_const
```

Out[27]:
```
array([[1.00000e+00, 5.80000e+01, 3.36400e+03, 1.95112e+05],
       [1.00000e+00, 5.90000e+01, 3.48100e+03, 2.05379e+05],
       [1.00000e+00, 6.00000e+01, 3.60000e+03, 2.16000e+05],
       [1.00000e+00, 6.10000e+01, 3.72100e+03, 2.26981e+05],
       [1.00000e+00, 6.20000e+01, 3.84400e+03, 2.38328e+05],
       [1.00000e+00, 6.30000e+01, 3.96900e+03, 2.50047e+05],
       [1.00000e+00, 6.40000e+01, 4.09600e+03, 2.62144e+05],
       [1.00000e+00, 6.50000e+01, 4.22500e+03, 2.74625e+05],
       [1.00000e+00, 6.60000e+01, 4.35600e+03, 2.87496e+05],
       [1.00000e+00, 6.70000e+01, 4.48900e+03, 3.00763e+05],
       [1.00000e+00, 6.80000e+01, 4.62400e+03, 3.14432e+05],
       [1.00000e+00, 6.90000e+01, 4.76100e+03, 3.28509e+05],
       [1.00000e+00, 7.00000e+01, 4.90000e+03, 3.43000e+05],
       [1.00000e+00, 7.10000e+01, 5.04100e+03, 3.57911e+05],
       [1.00000e+00, 7.20000e+01, 5.18400e+03, 3.73248e+05]])
```

42 统计分析

In [28]:
```
myModel_updated = sm.OLS(y, X_add_const)
```

In [29]:
```
results_updated = myModel_updated.fit()
print(results_updated.summary())
```

Out [29]:

```
                            OLS Regression Results
==============================================================================
Dep. Variable:                 weight   R-squared:                       1.000
Model:                            OLS   Adj. R-squared:                  1.000
Method:                 Least Squares   F-statistic:                 1.679e+04
Date:                Sun, 19 Aug 2018   Prob (F-statistic):           2.07e-20
Time:                        20:03:14   Log-Likelihood:                 1.3441
No. Observations:                  15   AIC:                             5.312
Df Residuals:                      11   BIC:                             8.144
Df Model:                           3
Covariance Type:            nonrobust
==============================================================================
                 coef    std err          t      P>|t|      [0.025      0.975]
------------------------------------------------------------------------------
const       -896.7476    294.575     -3.044      0.011   -1545.102    -248.393
x1            46.4108     13.655      3.399      0.006      16.356      76.466
x2            -0.7462      0.211     -3.544      0.005      -1.210      -0.283
x3             0.0043      0.001      3.940      0.002       0.002       0.007
==============================================================================
Omnibus:                        0.028   Durbin-Watson:                   2.388
Prob(Omnibus):                  0.986   Jarque-Bera (JB):                0.127
Skew:                           0.049   Prob(JB):                        0.939
Kurtosis:                       2.561   Cond. No.                     1.25e+09
==============================================================================
```

Warnings:
[1] Standard Errors assume that the covariance matrix of the errors is correctly specified.
[2] The condition number is large, 1.25e+09. This might indicate that there are strong multicollinearity or other numerical problems.

C:\Anaconda\lib\site-packages\scipy\stats\stats.py:1394: UserWarning: kurtosistest only valid for n>=20 ... continuing anyway, n=15"anyway, n=%i" % int(n))

In [30]:
```
print('查看斜率及截距项: ',results_updated.params)
```

查看斜率及截距项: const   -896.747633
x1   46.410789
x2   -0.746184
x3   0.004253
dtype: float64

In [31]:
 提示 | 重新预测体重。

```
y_predict_updated = results_updated.predict()
y_predict_updated
```

Out [31]: array([114.63856209, 117.40676937, 120.18801264, 123.00780722,
       125.89166846, 128.86511168, 131.95365223, 135.18280543,
       138.57808662, 142.16501113, 145.9690943,   150.01585147,
       154.33079796, 158.93944911, 163.86732026])

In [32]:
 提示 | 以下代码用于新结果的可视化。

```
plt.rcParams['font.family'] = "simHei"
```

 提示 | 上一行代码的主要作用是正常显示汉字。

```
plt.scatter(df_women["height"], df_women["weight"])
plt.plot(df_women["height"], y_predict_updated)
plt.title('女性身高与体重数据的线性回归分析')
plt.xlabel('身高')
plt.ylabel('体重')
```

Out [32]: Text(0,0.5,'体重')

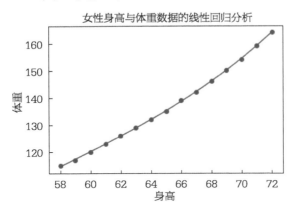

42.10 模型的应用

In [33]:

 提示 | 将模型应用于训练集和测试集之外的新数据,如预测身高为 63.5（英寸）的女性的体重（磅）。

h = 63.5
results_updated.predict([1,h,np.power(h,2),np.power(h,3)])

 注意 | 方法 predict() 的实参格式应与训练该模型时的自变量一致。读者可以通过 "results.predict?" 查看更多帮助信息。

Out [33]: array([130.39340008])

 扩展 | 相关资料。

- ❖ StatsModels 的官网：http://www.statsmodels.org/。
- ❖ Python 标准模块之"数值与数学模块"的官方文档：https://docs.python.org/3/library/numeric.html。
- ❖ 建议进一步深入学习数据分析/数据科学项目中常用的统计模型与机器学习算法，参见本书【50.4 常用统计模型（P495）】和【50.5 核心机器学习算法（P496）】。

43 机器学习

 常见疑问及解答。

- 常用Python机器学习包有哪些？
  - Scikit-Learn
  - Mlpy
  - TensorFlow
  - Keras/TensorFlow/Theano（深度学习）

- 如何拆分训练集和测试集？
  - train_test_split() —如→
    ```
    from sklearn.model_selection import train_test_split
    X_trainingSet, X_testSet, y_trainingSet, y_testSet=
    train_test_split(X_data, y_data, random_state=1)
    ```

- Python机器学习的一般步骤及方法是什么？
  - 1→ 业务理解 → 开展数据分析和数据科学项目的前提
  - 2→ 数据读入 —如→ `bc_data=pd.read_csv('bc_data.csv',header=0)`
  - 3→ 数据理解 —如→ `bc_data.shape` / `bc_data.describe()` / 数据可视化
  - 4→ 数据准备 —如→ 数据规整处理 / 训练集和测试集的拆分
  - 5→ 算法选择及其超级参数的设置 —如→ `myModel = KNeighborsClassifier(algorithm=kd_tree)`
  - 6→ 具体模型的训练 —如→ `myModel.fit(X_trainingSet, y_trainingSet)`
  - 7→ 用模型进行预测 —如→ `y_predictSet = myModel.predict(X_testSet)`
  - 8→ 模型评价 —如→ `from sklearn.metrics import accuracy_score` / `accuracy_score(y_testSet, y_predictSet)`
  - 9→ 模型的优化与应用 → 用模型进行预测新数据 / 优化模型或选择其他算法

43 机器学习

In [1]: 在数据分析和数据科学项目中,"训练集"与"测试集"是两个不同的概念,如下所示。

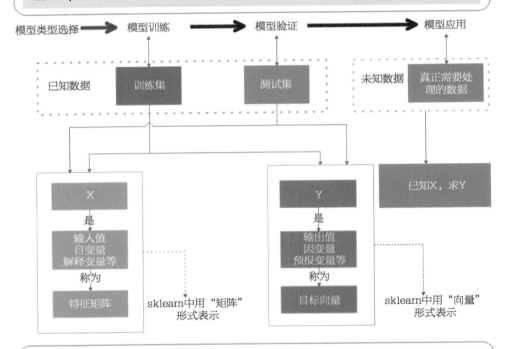

In [2]: 在数据分析和数据科学项目中,"特征矩阵"与"目标向量"是两个不同的概念,参见本书【42 统计分析(P370/In[1])】。

43.1 机器学习的业务理解

In [3]: "业务理解"是数据分析和数据科学项目的第一步。

 本例的主要业务场景如下。

#【数据及分析对象】
　　#CSV 文件——文件名为 bc_data.csv,数据内容来自"威斯康星乳腺癌

数据库（Wisconsin Breast Cancer Database）"
#ID：病例的 ID
#Diagnosis（医生给出的诊断结果）：M 为恶性，B 为良性。该数据集共包含 357 个良性病例和 212 个恶性病例
#细胞核的 10 个特征值：radius(半径)、texture(纹理)、perimeter（周长）、面积（area）、平滑度（smoothness）、紧凑度（compactness）、凹面（concavity）、凹点（concave points）、对称性（symmetry）和分形维数（fractal dimension）等。同时，为上述 10 个特征值均提供了三种统计量，分别为均值（mean）、标准差（standard error）和最大值（worst or largest）

#【目的及分析任务】
#理解机器学习方法在数据科学中的应用 KNN 方法进行分类分析
#首先，以随机选择的部分记录为训练集进行学习概念"诊断结果（diagnosis）"
#其次，以剩余记录为测试集，进行 KNN 建模
#接着，按 KNN 模型预测测试集的 diagnosis 类型
#最后，将 KNN 模型给出的 diagnosis "机器给出的预测诊断结果"与数据集 bc_data.csv 自带的"医生给出的诊断结果"进行对比分析，验证 KNN 建模的有效性

43.2 数据读入

In [4]:

 提示　建议先将准备读入的外部文件放在"当前工作目录"中，或者在读入外部文件之前设置"当前工作目录"。参见本书【34 当前工作目录（P224）】。

```
import pandas as pd
import numpy as np
import os
os.chdir(r'C:\Users\soloman\clm')
```

43 机器学习

 此处,"C:\Users\soloman\clm"是本书作者计算机的工作路径,建议读者根据自己计算机的情况设置具体工作路径。

print(os.getcwd())

Out [4]: C:\Users\soloman\clm

In [5]:

 读者可以从本书配套资源中找到数据文件 bc_data.csv。

bc_data = pd.read_csv('bc_data.csv', header = 0)

 数据文件 bc_data.csv 中的第 0 列为列名,所以 header 设置为 0。

bc_data.head()

 显示数据框的前 5 行,参见本书【38 DataFrame(P277/In[4])】。

Out [5]:

	id	diagnosis	radius_mean	texture_mean	perimeter_mean	area_mean	smooth
0	842302	M	17.99	10.38	122.80	1001.0	0.11840
1	842517	M	20.57	17.77	132.90	1326.0	0.08474
2	84300903	M	19.69	21.25	130.00	1203.0	0.10960
3	84348301	M	11.42	20.38	77.58	386.1	0.14250
4	84358402	M	20.29	14.34	135.10	1297.0	0.10030

5 rows × 32 columns

43.3 数据理解

In [6]:

 查看形状,参见本书【38 DataFrame(P279/In[9])】。

print(bc_data.shape)

Out [6]: (569, 32)

In [7]:
 查看列名，参见本书【38 DataFrame（P278/In[7]）】。

print(bc_data.columns)

Out [7]: Index(['id', 'diagnosis', 'radius_mean', 'texture_mean', 'perimeter_mean',
       'area_mean', 'smoothness_mean', 'compactness_mean', 'concavity_mean',
       'concave points_mean', 'symmetry_mean', 'fractal_dimension_mean',
       'radius_se', 'texture_se', 'perimeter_se', 'area_se', 'smoothness_se',
       'compactness_se', 'concavity_se', 'concave points_se', 'symmetry_se',
       'fractal_dimension_se', 'radius_worst', 'texture_worst',
       'perimeter_worst', 'area_worst', 'smoothness_worst',
       'compactness_worst', 'concavity_worst', 'concave_points_worst',
       'symmetry_worst', 'fractal_dimension_worst'], dtype = 'object')

In [8]:
 进行描述性统计分析，参见本书【38 DataFrame（P297/In[62]）】。

print(bc_data.describe())

Out [8]:

	id	radius_mean	texture_mean	perimeter_mean	area_mean\
count	5.690000e+02	569.000000	569.000000	569.000000	569.000000
mean	3.037183e+07	14.127292	19.289649	91.969033	654.889104
std	1.250206e+08	3.524049	4.301036	24.298981	351.914129
min	8.670000e+03	6.981000	9.710000	43.790000	143.500000
25%	8.692180e+05	11.700000	16.170000	75.170000	420.300000
50%	9.060240e+05	13.370000	18.840000	86.240000	551.100000
75%	8.813129e+06	15.780000	21.800000	104.100000	782.700000
max	9.113205e+08	28.110000	39.280000	188.500000	2501.000000

	smoothness_mean	compactness_mean	concavity_mean	Concave points_mean\
count	569.000000	569.000000	569.000000	569.000000
mean	0.096360	0.104341	0.088799	0.048919
std	0.014064	0.052813	0.079720	0.038803

min	0.052630	0.019380	0.000000	0.000000
25%	0.086370	0.064920	0.029560	0.020310
50%	0.095870	0.092630	0.061540	0.033500
75%	0.105300	0.130400	0.130700	0.074000
max	0.163400	0.345400	0.426800	0.201200

	symmetry_mean	...	radius_worst	texture_worst\
count	569.000000	...	569.000000	569.000000
mean	0.181162	...	16.269190	25.677223
std	0.027414	...	4.833242	6.146258
min	0.106000	...	7.930000	12.020000
25%	0.161900	...	13.010000	21.080000
50%	0.179200	...	14.970000	25.410000
75%	0.195700	...	18.790000	29.720000
max	0.304000	...	36.040000	49.540000

	Perimeter_worst	area_worst	smoothness_worst	Compactness_worst\
count	569.000000	569.000000	569.000000	569.000000
mean	107.261213	880.583128	0.132369	0.254265
std	33.602542	569.356993	0.022832	0.157336
min	50.410000	185.200000	0.071170	0.027290
25%	84.110000	515.300000	0.116600	0.147200
50%	97.660000	686.500000	0.131300	0.211900
75%	125.400000	1084.000000	0.146000	0.339100
max	251.200000	4254.000000	0.222600	1.058000

	concavity_worst	concave_points_worst	symmetry_worst\
count	569.000000	569.000000	569.000000
mean	0.272188	0.114606	0.290076
std	0.208624	0.065732	0.061867
min	0.000000	0.000000	0.156500
25%	0.114500	0.064930	0.250400
50%	0.226700	0.099930	0.282200
75%	0.382900	0.161400	0.317900
max	1.252000	0.291000	0.663800

	fractal_dimension_worst
count	569.000000
mean	0.083946
std	0.018061
min	0.055040
25%	0.071460
50%	0.080040
75%	0.092080
max	0.207500

[8 rows x 31 columns]

43.4 数据准备

In [9]:

 在数据科学中，数据准备的重点是数据的规整化处理。

```
#第一，数据加工：冗余数据、错误数据与缺失数据的处理
#第二，特征矩阵与目标向量的定义
#第三，测试数据与训练数据的拆分
```

In [10]:

 数据加工：本例中，删除没有实际意义的 ID 项数据，可以考虑直接删除它。

```
data = bc_data.drop(['id'], axis = 1)
print(data.head())
```

Out [10]:

	diagnosis	radius_mean	Texture_mean	perimeter_mean	area_mean\
0	M	17.99	10.38	122.80	1001.0
1	M	20.57	17.77	132.90	1326.0
2	M	19.69	21.25	130.00	1203.0
3	M	11.42	20.38	77.58	386.1
4	M	20.29	14.34	135.10	1297.0

	smoothness_mean	compactness_mean	concavity_mean	concave points_mean \
0	0.11840	0.27760	0.3001	0.14710
1	0.08474	0.07864	0.0869	0.07017
2	0.10960	0.15990	0.1974	0.12790
3	0.14250	0.28390	0.2414	0.10520
4	0.10030	0.13280	0.1980	0.10430

	symmetry_mean	...	radius_worst	texture_worst \
0	0.2419	...	25.38	17.33
1	0.1812	...	24.99	23.41
2	0.2069	...	23.57	25.53
3	0.2597	...	14.91	26.50
4	0.1809	...	22.54	16.67

	perimeter_worst	area_worst	smoothness_worst	compactness_worst \
0	184.60	2019.0	0.1622	0.6656
1	158.80	1956.0	0.1238	0.1866
2	152.50	1709.0	0.1444	0.4245
3	98.87	567.7	0.2098	0.8663
4	152.20	1575.0	0.1374	0.2050

	concavity_worst	concave_points_worst	symmetry_worst \
0	0.7119	0.2654	0.4601
1	0.2416	0.1860	0.2750
2	0.4504	0.2430	0.3613
3	0.6869	0.2575	0.6638
4	0.4000	0.1625	0.2364

	fractal_dimension_worst
0	0.11890
1	0.08902
2	0.08758
3	0.17300
4	0.07678

[5 rows x 31 columns]

In [11]

 定义特征矩阵。

X_data = data.drop(['diagnosis'], axis = 1)

 axis = 1 的含义为：行数不变，按行为单位计算，逐行计算。

X_data.head()

Out [11]:

	radius_mean	texture_mean	perimeter_mean	area_mean	smoothness_mean	compactn
0	17.99	10.38	122.80	1001.0	0.11840	0.27760
1	20.57	17.77	132.90	1326.0	0.08474	0.07864
2	19.69	21.25	130.00	1203.0	0.10960	0.15990
3	11.42	20.38	77.58	386.1	0.14250	0.28390
4	20.29	14.34	135.10	1297.0	0.10030	0.13280

5 rows × 30 columns

In [12]

 定义目标向量。

y_data = np.ravel(data[['diagnosis']])

 在数据分析和数据科学项目中，可以用内置函数 np.ravel() 进行降维处理。

y_data[0:6]

Out [12]: array(['M', 'M', 'M', 'M', 'M', 'M'], dtype = object)

In [13]

 测试数据与训练数据的拆分方法：用模块 sklearn.model_selection 中的 train_test_split()。

from sklearn.model_selection import train_test_split

```
X_trainingSet, X_testSet, y_trainingSet, y_testSet = train_test_split(X_data, y_data, random_state = 1)
```

 在上面的代码中，X_trainingSet 和 y_trainingSet 分别为训练集的特征矩阵和目标向量。

 X_testSet 和 y_testSet 分别为测试集的特征矩阵和目标向量。

In [14]:

 查看训练集的形状。

```
print(X_trainingSet.shape)
```

Out [14]: (426, 30)

In [15]:

查看测试集的形状。

```
print(X_testSet.shape)
```

Out [15]: (143, 30)

43.5 算法选择及其超级参数的设置

In [16]:

 （1）选择算法：本例选用 KNN，需要导入 KNeighborsClassifier 分类器。

```
from sklearn.neighbors import KNeighborsClassifier
```

In [17]:

（2）实例化 KNN 模型，并设置超级参数 algorithm = 'kd_tree'。scikit-learn 包提供了 KNN 近邻的三种查找方法：kd_tree、ball_tree 和 brute，详见其官方文档。

```
myModel = KNeighborsClassifier(algorithm = 'kd_tree')
```

43.6 具体模型的训练

In [18]:

> **提示** 基于"训练集"训练出新的具体模型。

myModel.fit(X_trainingSet, y_trainingSet)

> **提示** 训练集的特征矩阵:X_trainingSet。
> 训练集的目标向量:y_trainingSet。

Out [18]: KNeighborsClassifier(algorithm = 'kd_tree', leaf_size = 30, \
　　　　　meric = 'minkowski', metric_params = None, n_jobs = 1, \
　　　　　n_neighbors = 5, p = 2, weights = 'uniform')

43.7 用模型进行预测

In [19]:

> **思路** 将上一步中已训练出的具体模型用于测试集中的特征矩阵,预测对应的目标向量。

y_predictSet = myModel.predict(X_testSet)

> **提示** 测试集的特征矩阵:X_testSet。

In [20]:

> **提示** 查看预测结果。

print(y_predictSet)

43 机器学习

Out [20]: ['M' 'M' 'B' 'M' 'M' 'M' 'M' 'M' 'B' 'B' 'B' 'M' 'M' 'B' 'B' 'B' 'B'
'B' 'M' 'B' 'B' 'M' 'B' 'M' 'B' 'B' 'M' 'M' 'M' 'B' 'M' 'B' 'B' 'B'
'M' 'B' 'B' 'B' 'B' 'B' 'B' 'B' 'B' 'B' 'B' 'B' 'M' 'M' 'M' 'B'
'B' 'B' 'B' 'B' 'B' 'B' 'B' 'B' 'B' 'B' 'B' 'B' 'B' 'B' 'B' 'B'
'M' 'M' 'B' 'B' 'B' 'B' 'B' 'M' 'B' 'B' 'B' 'M' 'B' 'M' 'M' 'B'
'B' 'M' 'B' 'B' 'B' 'B' 'B' 'B' 'M' 'B' 'B' 'B' 'B' 'B' 'B' 'B'
'M' 'M' 'B' 'B' 'B' 'B' 'M' 'M' 'B' 'B' 'B' 'M' 'M' 'B' 'B' 'M'
'M' 'M' 'M' 'M' 'B' 'B' 'B' 'M' 'B' 'M' 'M' 'B' 'B' 'M' 'M' 'B']

In [21]:

 查看真实值。

```
print(y_testSet)
```

Out [21]: ['B' 'M' 'B' 'M' 'M' 'M' 'M' 'M' 'B' 'B' 'M' 'M' 'B' 'B' 'B' 'B' 'B'
'B' 'M' 'B' 'B' 'M' 'B' 'M' 'B' 'B' 'M' 'M' 'M' 'B' 'M' 'M' 'B' 'B'
'M' 'B' 'M' 'B' 'B' 'B' 'B' 'B' 'B' 'B' 'B' 'B' 'M' 'M' 'M' 'B'
'B' 'B' 'B' 'B' 'B' 'B' 'B' 'B' 'B' 'B' 'B' 'B' 'B' 'B' 'B' 'B'
'M' 'M' 'B' 'B' 'B' 'B' 'B' 'M' 'B' 'B' 'M' 'M' 'B' 'M' 'M' 'B'
'B' 'M' 'B' 'B' 'B' 'B' 'B' 'B' 'M' 'B' 'B' 'B' 'B' 'B' 'B' 'B'
'M' 'M' 'B' 'B' 'B' 'B' 'M' 'M' 'B' 'B' 'B' 'M' 'M' 'B' 'B' 'M'
'M' 'B' 'M' 'B' 'B' 'B' 'M' 'B' 'M' 'M' 'B' 'B' 'B' 'M' 'M' 'B']

43.8 模型评价

In [22]:

 导入 accaccuracy_score()函数，用于计算模型的准确率。

```
from sklearn.metrics import accuracy_score
```

 查看模型的准确率。

```
print(accuracy_score(y_testSet, y_predictSet))
```

 提示 y_testSet 和 y_predictSet 分别为测试集和预测集。

Out [22]: 0.9370629370629371

43.9 模型的应用与优化

In [23]:

 思路 采用绘制手肘曲线（Elbow Curve）选择 k 值。

```
from sklearn.neighbors import KNeighborsClassifier
NumberOfNeighbors = range(1,23)
KNNs = [KNeighborsClassifier(n_neighbors=i) for i in NumberOfNeighbors]
scores = [KNNs[i].fit(X_trainingSet, y_trainingSet).score(X_testSet, y_testSet) for i in range(len(KNNs))]
```

 提示 分别计算 k=1 到 k=23 时的 KNN 模型的准确率，并放在列表 scores 中。

```
scores
```

Out [23]: [0.9230769230769231,
0.9020979020979021,
0.9230769230769231,
0.9440559440559441,
0.9370629370629371,
0.9230769230769231,
0.9300699300699301,
0.9230769230769231,

0.9230769230769231,
0.9230769230769231,
0.9230769230769231,
0.9230769230769231,
0.9230769230769231,
0.9230769230769231,
0.9230769230769231,
0.916083916083916,
0.916083916083916,
0.916083916083916,
0.916083916083916,
0.916083916083916,
0.916083916083916,
0.9090909090909091]

In [24]:

 绘制手肘曲线（Elbow Curve）。

```
import matplotlib.pyplot as plt
%matplotlib inline
plt.plot(NumberOfNeighbors,scores)
plt.rcParams['font.family'] = 'simHei'
plt.xlabel('k 值')
plt.ylabel('得分')
plt.title('Elbow Curve')
plt.xticks(NumberOfNeighbors)
plt.show()
```

 从手肘曲线的显示结果可以看出，k=4 时准确率最高。

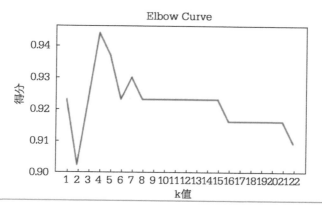

In [25]:

思路 将 k=4 代入 KNN 模型，进行重新预测。

```
from sklearn.neighbors import KNeighborsClassifier
myModel = KNeighborsClassifier(algorithm='kd_tree', n_neighbors=4)
myModel.fit(X_trainingSet, y_trainingSet)
y_predictSet = myModel.predict(X_testSet)
from sklearn.metrics import accuracy_score
print(accuracy_score(y_testSet, y_predictSet))
```

提示 准确率已提高至 0.9440559440559441。

Out [25]: 0.9440559440559441

In [26]:

思路 利用专用于机器学习的可视化 Python 包 Yellowbrick，绘制 ROC 曲线

```
from yellowbrick.classifier import ROCAUC
visualizer = ROCAUC(myModel, classes=["M", "B"])
```

提示 安装 Python 包 Yellowbrick 的命令为 pip install yellowbrick。更多内容见 Yellowbrick 包官网 https://www.scikit-yb.org/en/latest/api/cluster/elbow.html

43 机器学习

```
visualizer.fit(X_trainingSet, y_trainingSet)
visualizer.score(X_testSet, y_testSet)
visualizer.show()
```

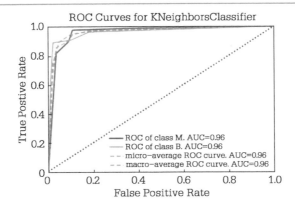

Out [26]:　　<AxesSubplot:title={'center':'ROC Curves for KNeighborsClassifier'}, xlabel='False Positive Rate', ylabel='True Postive Rate'>

 数据科学家是"科学家"吗?

数据科学家（Data Scientist）是一种新兴职业（Davenport T. H 和 Patil D. J 曾提出"数据科学家是 21 世纪最性感的职业"），不一定也不要求必须是传统意义上的"科学家"。当然，不排除有些数据科学家是真正的"科学家"。参见著名文章"Davenport T. H, Patil D. J. Data Scientist: The Sexiest Job of the 21st Century [J]. Harvard Business Review, 2012, 90(5): 70-76"。

44 自然语言处理

 常见疑问及解答。

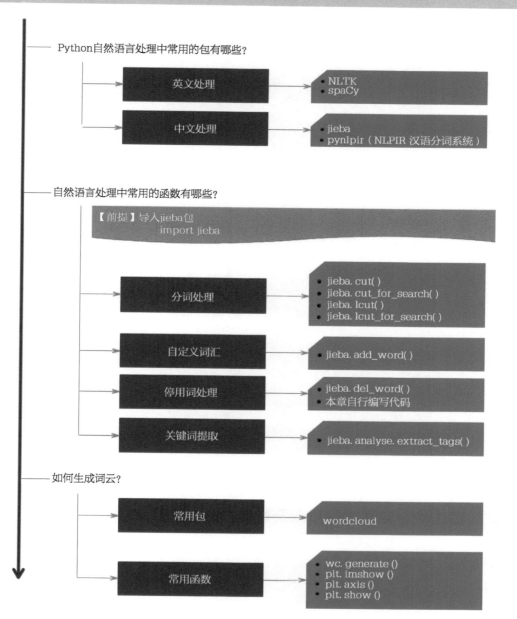

44 自然语言处理

44.1 常用包

In [1]:

 英文自然语言处理常用包：

#NLTK（Natural Language Toolkit）
#spaCy（Industrial-Strength Natural Language Processing）

 中文自然语言处理常用包：

#jieba：结巴分词工具
#pynlpir：NLPIR 汉语分词系统

In [2]:

 本书以中文自然语言处理工具 jieba 为例：

#对2021年央视春晚主持人主持词文本（文件名为2021.txt）进行文本分析
#分析任务为"分析2021年央视春晚主持人用词特点"
#读者可在本书配套资源中找到主持词文本文件（2021.txt）、停用词表
（stopwords.txt）以及Jieba词性中英文名称对照表（jiebaPOS.xlsx）

停用词表 jiebaPOS.xlsx 表为作者自定义表，读者可以视自己需要动态更新。

读者可在本书配套资源中可以找到本章所涉及的所有数据文件，建议将数据文件统一放在当前工作目录下的 data 文件夹中。

44.2 自然语言处理包的导入及设置

In [3]:

 jieba 包的参考资料

#https://github.com/fxsjy/jieba

In [4]:
> 本章需要安装的包及其安装方法如下：
>
> ```
> #!pip install jieba
> #!pip install WordCloud
> ```

In [5]:
> 包导入
>
> ```python
> import jieba
> import jieba.posseg as pseg    #词性标注器
>
> import numpy as np
> import pandas as pd
>
> import matplotlib as mpl
> import matplotlib.pyplot as plt
>
> %matplotlib inline
> from wordcloud import WordCloud
> ```

44.3 数据读入

In [6]:
> 读入主持词文本文件——2021.txt，建议读者将数据文件放在"当前目录"下的 data 文件夹中。
>
> ```
> text=open('data/2021.txt', 'r', encoding='utf-8').read().replace('\n','')
> ```
>
> 以只读方式（"r"）读入数据文件 2021.txt 至数据对象 text。同时，将数据文件 2021.txt 中的段落分割标志（"\n"）替换成空字符（""）。

实参"encoding='utf-8'"的含义为设置字符集。查看 Python 默认字符集的方法如下：

```
#import sys
#sys.getdefaultencoding()
```

 显示函数 open() 的返回值 text 的前 500 个字符

```
text[:500]
```

Out [6]: '任鲁豫：玉鼠追冬去，金牛送春来。尼格买提：全国和全世界的观众、听众朋友们，随着辛丑牛年的款款来临，中国中央广播电视总台 2021 年春节联欢晚会在这里和您见面啦！李思思：即将辞别的旧岁，极不平凡。我们在风雨中前行，经历了太多太多。张韬：这一年，我们哭过、笑过、拼过，这一年，我们每个人都了不起。龙洋：浮云难蔽日，雾散终有时。任鲁豫：今晚在这阖家团圆、辞旧迎新的时刻，我们要向所有的中国人，深情地道一声，你们合：辛苦了。李思思：朋友们，我们的晚会正在通过央视综合频道、综艺频道、中文国际频道、国防军事频道、少儿频道、农业农村频道、4K 超高清频道和 8K 超高清试验频道，以及央广音乐之声、经典音乐广播、文艺之声、中国交通广播、华语环球广播和大湾区之声、南海之声等同步直播。尼格买提：与此同时，央视频、央视新闻新媒体、央视网、央广网、国际在线、云听等新媒体平台同步播出。总台英语、西班牙语、法语、阿拉伯语、俄语频道和 43 种外语新媒体将联动全球 170 多个国家和地区的 600 多家媒体对春晚进行直播和报道。龙洋：通过这些传播平台，我们晚会的盛况将在同一时刻传遍神州大地，传遍五洲四海。张韬：通过这些传播平台，我们要向全'

In [7]: open() 函数的返回值为字符串（str）。

```
type(text)
```

Out [7]: str

In [8]:
```
#open?
```

44.4 分词处理

In [9]:

 在正式分词处理之前，先进行探索性分词。目前，jieba 支持四种分词模式：

精确模式，试图将句子最精确地切开，适合文本分析；
全模式，把句子中所有可以成词的词语都扫描出来，速度快，但是不能解决歧义；
搜索引擎模式，在精确模式的基础上，对长词再次切分，提高召回率，适合用于搜索引擎分词；
paddle 模式，利用 PaddlePaddle 深度学习框架，训练序列标注（双向 GRU）网络模型实现分词，同时支持词性标注。

words = pseg.cut(text[:20]) #采用精确模式分词#

 考虑到显示篇幅，在此仅对 text 的前 20 个字符进行分词，建议读者将代码"text[:20]"改为"text"。

for word, flag in words:
    print(F'{word}　{flag}') #print函数的实参为F-String

 通过探索性分词，发现如下 3 种问题，需要进一步优化分词方案。

#（1）分词问题：某些词的切分不准确，如"央视"等，需要自定义词汇
#（2）停用词问题：需要排除掉个别词语，如"主持人："，需要处理停用词
#（3）缺"年份"标签

Building prefix dict from the default dictionary ...
Loading model from cache C:\Users\chaol\AppData\Local\Temp\jieba.cache
Loading model cost 0.684 seconds.
Prefix dict has been built successfully.
任鲁豫　nr

44.5 自定义词汇

In [10]:

> **提示** 自定义词汇的方法：jieba.add_word(word, freq=None, tag=None)，其参数含义如下：

```
：    x
玉鼠   n
追    v
冬去   t
，    x
金
牛    nz
送    v
春
来    t
。    x
尼格   nrt
买    v
提    v
```

```
#word（必选）为新增词汇
#freq（可选）为出现频率
#tag（可选）为词性
jieba.add_word("央视",tag="n")
jieba.add_word('中国中央广播电视总台',tag="n")
jieba.add_word('抖音APP',tag="n")
jieba.add_word('抖音平台',tag="n")
jieba.add_word('尼格买提',tag="n")
jieba.add_word('任鲁豫',tag="n")
jieba.add_word('李思思',tag="n")
jieba.add_word('张韬',tag="n")
jieba.add_word('龙洋',tag="n")
jieba.add_word('合：',tag="n")
jieba.add_word('朋友们',tag="n")
jieba.add_word('海澜之家',tag="n")
jieba.add_word('张梓琳',tag="n")
jieba.add_word('张家丰',tag="n")
```

In [11]:

技巧 | 查看 jieba.add_word()函数的帮助信息的方法：

```
help(jieba.add_word)
```

Help on method add_word in module jieba:

add_word(word, freq=None, tag=None) method of jieba.Tokenizer instance
    Add a word to dictionary.

    freq and tag can be omitted, freq defaults to be a calculated value
    that ensures the word can be cut out.

In [12]:

提示 | 自定义词汇之后，再次对数据进行重新分词，查看是否需要继续更新自定义词汇表：

```
words = pseg.cut(text[:20])
for word, flag in words:
    print(F"{word}    {flag}") #print 函数的实参为 F-String
```

```
任鲁豫    n
:       x
玉鼠     n
追      v
冬去     t
,       x
金牛     nz
送      v
春来     t
。      x
尼格买提   n
```

In [13]:

技巧 | 查看 jieba 分词器的帮助信息

```
help(pseg.cut)
```

Help on function cut in module jieba.posseg:

cut(sentence, HMM=True, use_paddle=False)

Global `cut` function that supports parallel processing.

Note that this only works using dt, custom POSTokenizer instances are not supported.

In [14]:

 对 2021 年春晚主持词数据进行分词，并追加至一个新的列表 words 中。

```
words = [] #定义空列表

year=2021
year_words = [] #定义空列表

year_words.extend(pseg.cut(text))
```

 此处，year_words 的数据类型为列表，但是其中所包含的每一个元素的数据类型为元组，#即 years_words 的内容为 [('主持人：', '名词'), ('中国', '名词'),…]。Python 中的元组是一种不可变对象，无法修改，需要转换成列表并在词性标注中增加一个名为"年份"的新列，参见本书【15 元组（P97/In[7]）】

```
for j in range(len(year_words)):
    ls_year_words=list(year_words[j])
    ls_year_words.append(year)
    words.append(ls_year_words)
```

注意 .extend()方法与.append()方法不同，参见本书【14 列表（P84/In[25]）】。

```
words[0:13]
```

Out [14]: [['任鲁豫', 'n', 2021],
['：', 'x', 2021],
['玉鼠', 'n', 2021],
['追', 'v', 2021],
['冬去', 't', 2021],
['，', 'x', 2021],
['金牛', 'nz', 2021],
['送', 'v', 2021],

```
['春来', 't', 2021],
['。', 'x', 2021],
['尼格买提', 'n', 2021],
[': ', 'x', 2021],
['全国', 'n', 2021]]
```

In [15]: 将列表 words 转换为数据框对象 df_words，并设置列名依次为"词汇""词性"和"年份"。

```
df_words = pd.DataFrame(words,columns=["词汇","词性","年份"])
df_words
```

Out [15]:

	词汇	词性	年份
0	任鲁豫	n	2021
1	:	x	2021
2	玉鼠	n	2021
3	追	v	2021
4	冬去	t	2021
...	...	...	...
6180	合	v	2021
6181	:	x	2021
6182	再见	v	2021
6183	!	x	2021
6184		x	2021

6185 rows × 3 columns

In [16]: 查看行数。

```
df_words.index.size
```

 计算行数的方法有多种，如 df_words.shape[0]、df_words.shape、df_words.info 等。

Out [16]: 6185

In [17]: 增加词性的中文名称

44 自然语言处理

```python
import pandas as pd
jiebapos=pd.read_excel("data\\jiebaPOS.xlsx",header=0)

df_words_renamed=df_words.join(jiebapos.set_index('词性英文名称'),
on='词性')
```

 函数 df_words.join（ ）的参数及其含义如下：

```
#jiebapos.set_index('词性英文名称')为数据框jiebapos设置index
#on='词性' 指的是df_words的'词性'一列

df_words_renamed
```

Out [17]:

	词汇	词性	年份	词性中文名称
0	任鲁豫	n	2021	名词
1	：	x	2021	非语素词
2	玉鼠	n	2021	名词
3	追	v	2021	动词
4	冬去	t	2021	时间词
...	...	...	...	...
6180	合	v	2021	动词
6181	：	x	2021	非语素词
6182	再见	v	2021	动词
6183	！	x	2021	非语素词
6184		x	2021	非语素词

6185 rows × 4 columns

In [18]

 查看"词性中文名称"一列中是否存在缺失值。当本项目提供的中英文词性对照表 jiebaPOS.xlsx 的内容与 jieba 包的最新词性列表不完全一致时，会导致 df_words_renamed 的【词性中文名称】一列中出现缺失值。

```
#如有缺失值，需要在【jiebaPOS.xlsx】增加对应的中文名称
df_words_renamed[df_words_renamed.词性中文名称.isnull()]
```

Out [18]: 词汇　词性　年份　词性中文名称

44.6 停用词处理

In [19]:
> 提示 读入停用词表 stopwords.txt，读者可以从本书配套资源中找到该文件。

```
stopwords= open(r'data/stopwords.txt').read()
lst_StopWords=stopwords.split("\n")
lst_StopWords[:10]
```

Out [19]: ['龙洋', '张韬', '任鲁豫', '李思思', '尼格买提', '主持人', '主持人：', '主持词', '《', '》']

In [20]:
> 提示 过滤停用词

```
df_words = df_words_renamed[df_words_renamed.apply(lambda x: x.loc["词汇"] not in lst_StopWords,axis=1)]
    #apply（）方法
    #axis=1的含义
df_words[:10] #查看停用词过滤后的数据框df_words的前10行。
```

Out [20]:

	词汇	词性	年份	词性中文名称
2	玉鼠	n	2021	名词
3	追	v	2021	动词
4	冬去	t	2021	时间词
6	金牛	nz	2021	其他专名
7	送	v	2021	动词
8	春来	t	2021	时间词
12	全国	n	2021	名词
14	全世界	n	2021	名词
16	观众	n	2021	名词
18	听众	n	2021	名词

In [21]:
> 提示 查看停用词处理后的词数

```
df_words.shape[0]
```

Out [21]: 4087

44.7 词性分布分析

In [22]:

 创建新数据框 df_WordSpeechDistribution,并统计 2021 央视春晚主持词的词性分布。

df_WordSpeechDistribution = pd. DataFrame(df_words['词性中文名称']. value_counts(ascending=False))

 value_counts()的主要功能:计数并按降序排序(ascending=False)

df_WordSpeechDistribution.head(10)

Out [22]:

	词性中文名称
名词	980
动词	787
数词	351
代词	331
副词	262
人名	204
形容词	150
时间词	144
介词	136
连词	86

In [23]:

 修改列名,将原"词性"列的名称改为"频数"

df_WordSpeechDistribution.rename(columns={'词性中文名称':'频数'},inplace=True)

 rename()函数的参数如下。

#inplace:是否修改数据框本身
#columns = {"旧名称 1": "新名称 1", "旧名称 2": "新名称 2"}

注意 | rename()与reindex()不同,后者不改变名称,只调整显示顺序(隐式索引)。

df_WordSpeechDistribution.head()

Out [23]:

	频数
名词	980
动词	787
数词	351
代词	331
副词	262

In [24]: df_WordSpeechDistribution['频数'].sum()

 技巧 | 查看行数,注意是否等于df_words.shape[0]。

Out [24]: 4087

In [25]: 提示 | 增设一个新列——"百分比"。

df_WordSpeechDistribution['百分比'] = df_WordSpeechDistribution['频数'] / df_WordSpeechDistribution['频数']. sum()
df_WordSpeedf_WordSpeechDistribution.head(10)

Out [25]:

	频数	百分比
名词	980	0.239785
动词	787	0.192562
数词	351	0.085882
代词	331	0.080989
副词	262	0.064106
人名	204	0.049914
形容词	150	0.036702
时间词	144	0.035234

44 自然语言处理

| 介词 | 136 | 0.033276 |
| 连词 | 86 | 0.021042 |

In [26]:

 提示 | 用 Pandas 的 plot()方法绘制前 10 位词性类别分布图。

```
plt.rcParams["font.family"] = 'simHei'
plt.subplots(figsize=(7,5))
plt.rcParams["font.family"] = 'STFangsong'
df_WordSpeechDistribution.iloc[:10]['频数'].plot(kind='barh')
plt.yticks(size=10)
plt.xlabel('频数',size=10)
plt.ylabel('词性',size=10)
plt.title('2021央视春晚主持人【主持词】词性分布分析')
```

Out[26]: Text(0.5,1,'2021央视春晚主持人"主持词"词性分布分析')

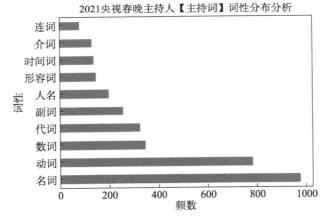

44.8 高频词分析

In [27]: df_words.head(20)

Out [27]:

	词汇	词性	年份	词性中文名称
2	玉鼠	n	2021	名词
3	追	v	2021	动词
4	冬去	t	2021	时间词
6	金牛	nz	2021	其他专名
7	送	v	2021	动词

8	春来	t	2021	时间词
12	全国	n	2021	名词
14	全世界	n	2021	名词
16	观众	n	2021	名词
18	听众	n	2021	名词
19	朋友们	n	2021	名词
21	随着	p	2021	介词
22	辛丑	nr	2021	人名
23	牛年	t	2021	时间词
25	款款	n	2021	名词
26	来临	v	2021	动词
28	中国中央广播电视总台	n	2021	名词
29	2021	m	2021	数词
30	年	m	2021	数词
31	春节	t	2021	时间词

In [28]:

 创建列表 columns_slected，并定义 6 个主要词性列表（除"人名"外）。

columns_slected=['名词','名词计数','动词','动词计数','数词','数词计数','代词','代词计数','副词','副词计数','形容词','形容词计数']

 定义 6 类词性统计数据框 df_Top6。

df_Top6 = pd.DataFrame(columns=columns_slected)

 统计 6 大词性类别的具体内容，并保存到数据 df_Top6 框中。其中，reset_index()的功能为将 Pandas 的 Series 改为数据框（Data Frame）。

```
for i in range(0,12,2):
    df_Top6[columns_slected[i]] = df_words.loc[df_words['词性中文名称']==columns_slected[i]]['词汇'].value_counts().reset_index()['index']
    df_Top6[columns_slected[i+1]] = df_words.loc[df_words['词性中文名称']==columns_slected[i]]['词汇'].value_counts().reset_index()['词汇']
```

44 自然语言处理

 查看数据框 df_Top6 的前 10 行。

```
df_Top6.head(10)
```

Out [28]:

	名词	名词计数	动词	动词计数	数词	数词计数	代词	代词计数	副词	副词计数	形容词	形容词计数
0	朋友们	41	要	28.0	年	29.0	我们	118.0	就	24.0	亲爱	17.0
1	红包	32	到	19.0	亿	16.0	您	31.0	将	17.0	好	17.0
2	大家	32	打开	19.0	一个	16.0	这	31.0	都	16.0	新	17.0
3	抖音APP	19	来	16.0	一年	14.0	他们	27.0	还	13.0	美好	5.0
4	掌声	17	说	15.0	一起	13.0	你	12.0	就是	9.0	伟大	4.0
5	牛	16	请	14.0	一	11.0	那	11.0	又	9.0	欢乐	3.0
6	人	15	下载	12.0	2021	7.0	你们	8.0	更	7.0	最美	3.0
7	现金	13	继续	11.0	12	7.0	我	8.0	不	7.0	健康	3.0
8	祖国	10	拜年	11.0	元	7.0	他	7.0	一直	7.0	幸福	3.0
9	戏曲	10	抖	10.0	5	6.0	其中	7.0	已经	7.0	平凡	3.0

44.9 词频统计

In [29]:

 查看 df_words 的当前值

```
df_words.head()
```

Out [29]:

	词汇	词性	年份	词性中文名称
2	玉鼠	n	2021	名词
3	追	v	2021	动词
4	冬去	t	2021	时间词
6	金牛	nz	2021	其他专名
7	送	v	2021	动词

In [30]:

> **提示** 只保留 df_words 的 "年份" 和 "词汇" 两列，并转换格式成数据框对象 df_AnnaulWords。其中，方法.pivot()的功能为生成新的透视表。

df_AnnualTopWords=pd.DataFrame(columns=[2021])

df_AnnualTopWords[2021]=df_words["词汇"].value_counts().reset_index()["index"]

> **提示** 最后加一个["index"]的目的是，"只保留 index 部分，因为还有一个频次列"。

df_AnnualTopWords[0:].head(10)

Out [28]:

	2021
0	我们
1	您
2	朋友们
3	谢谢
4	为
5	红包
6	大家
7	这
8	接下来
9	年

44.10 关键词分析

In [31]:

> **思路** 调用 jieba 关键词抽取模块 jieba.analyse

import jieba.analyse as analyse

> **提示** 定义一个数据框对象 df_annual_keywords，用于保存关键词提取结果。

df_annual_keywords = pd.DataFrame(columns=[2021])

44 自然语言处理

 逐年提取关键词，并保存到数据框中

```
df_annual_keywords[2021]=analyse.extract_tags(' '.join(df_AnnualTopWords[2021].astype('str')))
```

 jieba 提供两种关键词提取方法：TF-IDF 算法和 TextRank 算法。在此，.extract_tags()函数的功能为"基于 TF-IDF 算法的关键词抽取"，详细内容建议读者参考 jieba 官方文档。

 查看年度前 10 关键词

```
df_annual_keywords.head(10)
```

Out [31]:

	2021
0	新春快乐
1	扬帆远航
2	王奔
3	北京服装学院
4	孺子牛
5	好戏连台
6	五洲四海
7	叶蓬
8	吴京
9	再立新功

44.11 生成词云

In [32]:

 字符串 myText 的当前值（考虑到篇幅限制，在此仅显示前 200 个字符）

```python
myText=' '.join(df_words.词汇)
myText[:200]
```

Out [32]: '玉鼠 追 冬去 金牛 送 春来 全国 全世界 观众 听众 朋友们 随着 辛丑 牛年 款款 来临 中国中央广播电视总台 2021 年 春节 联欢晚会 这里 您 见面 啦 即 将 辞别 旧岁 极 不 平凡 我们 风雨 中 前行 经历 太多太多 这 一年 我们 哭 过 笑 过 拼过 这 一年 我们 每个 人 都 了不起 浮云 难 蔽日 雾 散 终 有时 今晚 这 阖家团圆 辞旧迎新 时刻 我们 要 向 所'

In [33]:

 提示 | 读入词云的背景图片

```python
from imageio import imread
bg_pic = imread('data/host.jpg')
```

In [34]:

 提示 | 下载、安装并导入包 wordcloud。下载和安装方法为：pip install wordcloud 命令。

 注意 | 在个别读者的计算机上，wordcloud 的安装可能需要下载 Microsoft Visual C++14.0，下载 URL 为 http://land-inghub.visuals tudio.com/visual-cpp-build-tools，选择择 StandaloneCompiler 即可。

生成要显示的文字wc.generate(text)中需要字符串，不需要带【频次】。

```python
from wordcloud import WordCloud
```

 提示 | 设置词云字体，读者需要视自己机器上的中文字体文件名称及路径修改font_wc的值。

```python
font_wc= r'C:\Windows\Fonts\msyhbd.ttc'
```

 提示 | 生成词云，并设置词云属性，包括字体、背景图片、背景颜色、最大词数、字体最大值、图片默认大小等。

```python
wc = WordCloud(font_path=font_wc
              , mask=bg_pic
              ,max_words=500
              ,max_font_size=200
              ,background_color='white'
              ,colormap= 'Reds_r'
              ,scale=15.5)
```

44 自然语言处理

In [35]: 查看 WordCloud()函数的帮助信息。

```
WordCloud?
```

In [36]: 生成词云——wc.generate();显示词云，plt.imshow();关闭/不显示 Matplotlib 的 X 和 Y 轴，plt.axis('off')。

```
wc.generate(myText)
plt.imshow(wc)
plt.axis('off')
```

Out [36]: (−0.5, 10353.5, 15499.5, −0.5)

In [37]: 查看函数 plt.imshow()的帮助信息

```
help(plt.imshow)
```

In [38]: 导出词云对象 dc 至 data 文件夹中，文件名为 wordCloud_SF_CCTV.jpg。导出方法：.to_file()

```
wc.to_file('data\\wordCloud_SF_CCTV.jpg')
```

Out [38]: <wordcloud.wordcloud.WordCloud at 0x2e7740d3970>

45 人脸识别与图像分析

 常见疑问及解答。

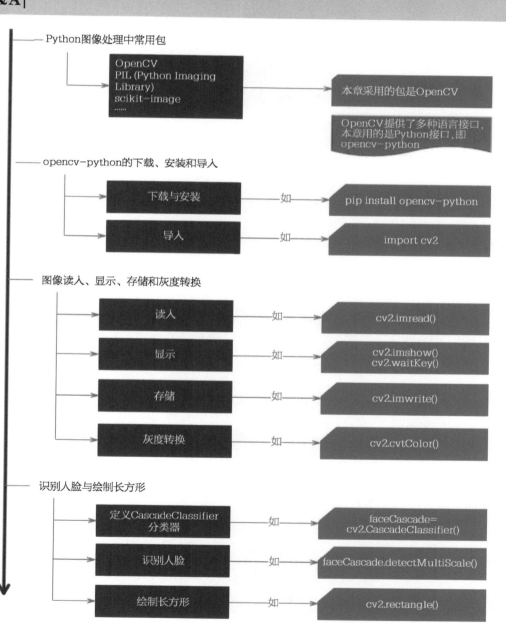

45 人脸识别与图像分析

45.1 安装并导入 opencv-python 包

In [1]:

思路 | 下载 opencv-python 的方法为：pip install opencv-python。若 pip 默认服务器下载速度太慢，则建议改为国内镜像站点。

提示 | OpenCV 是一种常用的跨平台计算机视觉与机器学习工具库，目前提供了数百个机器视觉工具包，本例采用其中的物体检测包 haarcascades 进行人脸识别。

思路 | 导入 opencv-python 包。

import cv2

45.2 读取图像文件

In [2]:

思路 | 调用 opencv-python 包中的 imread()方法，读取图像文件 test.jpg 至图像对象 image。

image = cv2.imread("images/test.jpg")

提示 | 读者可以在本书配套资料中找到文件 test.jpg

45.3 将 RGB 图像转换为灰度图

In [3]:

思路 | 即将调用的物体检测包 haarcascades 中人脸识别函数 detectMultiScale()的参数为灰度图，在此将彩色图像 image 转换为灰度图对象 gray。

gray = cv2.cvtColor(image, cv2.COLOR_BGR2GRAY)

调用 pencv-python 包中的 imshow()方法和 cv2.waitKey()方法，显示新生成的灰度图对象 gray

```
cv2.imshow("Showing gray image", gray)
```

 cv2.imshow()的参数"Showing gray image"和 gray 分别为显示窗口的名称和窗口显示的图像。

```
cv2.waitKey(0)
```

 函数 waitKey( )的功能为设置图像窗口的显示时长，waitKey(0)的含义为一直显示图像窗口，直至用户关闭窗口为止。

 显示结果如下所示。

Out [3]: -1

45.4 人脸识别与绘制长方形

In [4]: 调用 opencv-python 包中的层叠分类器（CascadeClassifier）提供的方法 detectMultiScale()检测图像中的人脸。

45 人脸识别与图像分析

 提示 CascadeClassifier 是 OpenCV 中常用的物体识别与检测工具包。

```
faceCascade=cv2.CascadeClassifier(cv2.data.haarcascades +
"haarcascade_frontalface_default.xml")
```

 提示 CascadeClassifier 提供了不同物体的识别模板，模板内容以 XML 文件形式存储。

#本例采用的模板文件为haarcascade_frontalface_default.xml，更多模板请参考其官网https://github.com/opencv/opencv/tree/master/data/haarcascades

 思路 调用 CascadeClassifier 提供的方法 detectMultiScale()，检测灰度图 gray 中的人脸。

```
faces=faceCascade.detectMultiScale(gray,scaleFactor=1.1,min-
Neighbors=5,minSize=(30,30))
```

 提示 关键词参数 scaleFactor、minNeighbors 和 minSize 的含义为每轮检测图像窗口缩放比例、最小被检测到几次才能判定对象确实存在和检测对象的最小尺寸。

 思路 调用 CascadeClassifier 提供的方法 rectangle()，在本立体原图 image 上为每个已检测到的人脸绘制长方形边框。

```
for (x,y,w,h) in faces:
    cv2.rectangle(image,(x,y),(x+w,y+h),(0,0,255),8)
```

 提示 此处，cv2.rectangle()的四个参数(x,y)、(x+w,y+h)、(0,0,255)和 8，分别为正方形边框的左上角的坐标（起始位置）、右下角的坐标（结束位置）、颜色和粗细。

45.5 图像显示

In [5]: **思路** 调用 opencv-python 包中的方法 imshow()，显示已人脸检测，并添加长方形边框的图像对象 image。

```
cv2.imshow("Window Name", image)
cv2.waitKey(0)
```

Out [5]: −1

45.6 图像保存

In [6]: 思路 | 将用矩阵标识后的图像输出到当前工作目录下的 images 目录，文件名为 test_fr.png。

```
cv2.imwrite("images/test_fr.png", image)
```

Out [6]: True

第六篇
大数据处理

Spark 编程

基于Spark和MongoDB的大数据分析

46 Spark 编程

 常见疑问及解答。

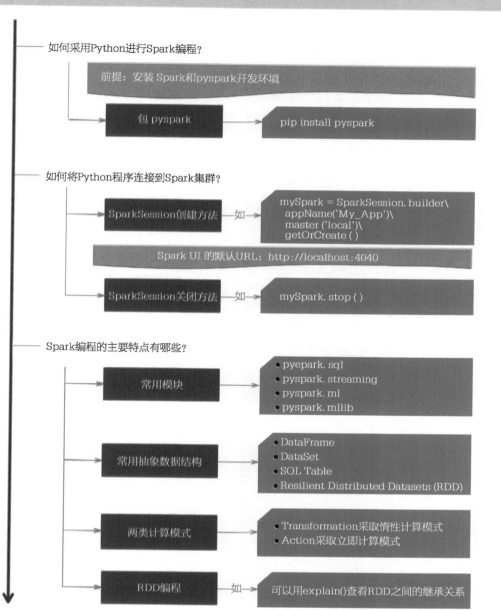

46 Spark 编程

如何进行Spark DataFrame的编程?

Spark DataFrame是在Spark平台上重新实现的、一种类似Pandas的DataFrame数据结构,支持大数据处理及Spark提供的API库

- 创建 Spark DataFrame —— 如 → df = mySpark.read.csv()
- 显示模式信息 —— 如 → df.printSchema()
- 进行缓存处理 —— 如 → df.cache()
- 显示内容 —— 如 → df.show()
- 显示列名 —— 如 → df.columns
- 统计行数 —— 如 → df.count()
- 选择特定行 —— 如 → df.select()
- 转换为临时视图 —— 如 → df.createTempView()
- 转换为本地 DataFrame —— 如 → df.toPandas()
- 导入/导出 —— 如 → df.write.csv()
  df.read.csv()

如何进行Spark SQL编程?

- SQL查询 → mySpark.sql(SQL查询语句)
- 过滤行 → df.filter(过滤条件)
- 分组统计 → df.groupBy().agg()

46.1 下载 pyspark 包

In [1]:
> 提示　在导入 pyspark 包前需要下载和安装它，方法为：在 CMD 窗口中输入 "pip install pyspark"。参见本书【26 模块（P169）】和【27 包（P175/In[2]）】。

46.2 SparkSession 及其创建

In [2]:
> 提示　（1）SparkSession 的重要地位：Python 代码与 Spark 集群的桥梁。
> #Spark 2.0 之前的版本，用户需要通过 SparkConf、Spark Context 和 SQLContext 等连接到 Spark 集群。
> #从 Spark 2.0 开始，SparkConf、Spark Context 和 SQLContext 被统一封装为 SparkSession。

```
from pyspark.sql import SparkSession
```

In [3]:
> 提示　（2）查看 Spark Session 帮助信息。

```
SparkSession.__doc__
```

Out [3]: 'The entry point to programming Spark with the Dataset and DataFrame API.\n\n A SparkSession can be used create: class:`DataFrame`, register :class:`DataFrame` as\n tables, execute SQL over tables, cache tables, and read parquet files.\n To create a SparkSession, use the following builder pattern:\n\n >>> spark = SparkSession.builder \\\n ... .master("local") \\\n ... .appName("Word Count") \\\n ... .config("sp ark.some.config.option", "some-value") \\\n ... .getOrCreate()\n '

In [4]:
```
dir(SparkSession)
```

Out [4]: ['Builder',

46 Spark 编程

```
'__class__',
'__delattr__',
'__dict__',
'__dir__',
'__doc__',
'__enter__',
'__eq__',
'__exit__',
'__format__',
'__ge__',
'__getattribute__',
'__gt__',
'__hash__',
'__init__',
'__init_subclass__',
'__le__',
'__lt__',
'__module__',
'__ne__',
'__new__',
'__reduce__',
'__reduce_ex__',
'__repr__',
'__setattr__',
'__sizeof__',
'__str__',
'__subclasshook__',
'__weakref__',
'_createFromLocal',
'_createFromRDD',
'_inferSchema',
'_inferSchemaFromList',
'_instantiatedSession',
'_repr_html_',
'builder',
'catalog',
```

'conf',
'createDataFrame',
'newSession',
'range',
'read',
'readStream',
'sparkContext',
'sql',
'stop',
'streams',
'table',
'udf',
'version']

In [5]:

 提示 可以用 help(SparkSession)方式查看帮助信息。

In [6]:

 提示 （3）创建 SparkSession 对象的方法：
SparkSession.builder()

 注意 本书中的 SparkSession 实例名称为 mySpark，用于连接至 Spark 集群。

```
mySpark = SparkSession.builder\
.appName('My_App')\
.master('local')\
.getOrCreate()
```

 注意 App 名称中不要带空格，否则会出错。别忘记续行符"\"，否则提示"Indentation Error: unexpected indent"。

 提示 "master('local')"的含义：以本地模式加载集群。
Spark UI 的链接为 http://localhost:4040/jobs/。

46 Spark 编程

In [7]: （4）查看 SparkSession 的实例 mySpark。

mySpark

Out [7]: SparkSession - in-memory
SparkContext

Spark UI
Version
v2.2.1
Master
local
AppName
My_App

46.3 Spark 数据抽象类型

In [8]: Spark 中的数据抽象类型如下。

#DataFrame
#Dataset
#SQL Table
#Resilient Distributed Datasets（RDD）。

In [9]: 以 DataFrame 为例，可以调用方法 toDF() 创建一个数据框。

myDF = mySpark.range(1,100).toDF("number")

 此处，range() 和 toDF() 均为 SparkSession 的方法，而不是 Pandas 或 Python 基础语法中的函数。

提示 toDF()方法用于将其他类型的数据转换为 Spark 数据框，实参"number"为目标列名。

In [10]:
 思路 Spark 特点之一，采用 Lazy Evaluation 思想，即将 Spark 操作分为 Transformation 和 Action 两大类操作。

print(myDF)

 提示 Spark 中的 print()函数并没有输出结果，仅仅输出了模式信息，如 DataFrame[number: bigint]。原因分析：Spark 采用的是"惰性计算（Lazy Conputing）"技术。在此，range()和 toDF()方法均采取"惰性计算"而不是"立即计算"。

DataFrame[number: bigint]

In [11]:
 提示 显示模式的方法：
printSchema()

myDF.printSchema()

root
 |-- number: long (nullable = false)

In [12]:
 思路 Spark 特点之二，Transformation 类操作的主要特征。

#Spark 的基本数据结构都是不可更改的（immutable），即一旦生成后不得修改它，而其修改操作是通过 Transformation（转换）来实现的

In [13]:
 提示 在 Spark 中，Transformation 只做了如下两件事情。

#（1）制订了一个操作规划（plan）
　先不执行，等到运行一个 Action（行动）中需要相应 Transformation（转换）的结果时，返过来执行 Transformation（转换）
#（2）定义了模式

46 Spark 编程

因此，以下代码没有立即显示最终结果，而是只显示输出结果的模式信息。

```
divisBy2 = myDF.where("number % 2 = 0")
divisBy2
```

Out [13]: DataFrame[number: bigint]

In [14]:

Spark 特点之三，Action（行动）类操作的主要特征。

```
#与 Transformation 不同的是，Action 是立即计算
divisBy2.count()
```

Action 会被立即计算，并生成计算结果。

Out [14]: 49

In [15]:
```
myDF.take(5)
```

显示前 5 行，与 myDF.head(5) 相似。

Out[15]: [Row(number = 1), Row(number = 2), Row(number = 3), Row(number = 4), Row(number = 5)]

In [16]:

查看 Spark UI 的方法如下。

```
#在浏览器中输入 http://localhost:4040
```

In [17]:

如何记录基于Transformation计算结果之间的继承关系呢？用RDD。

```
#RDD 就是一个 Physical Plan（物理规划）
myDF = mySpark.range(1,100).toDF("number").where("number % 2 = 0").sort("number")
myDF
```

Out [17]: DataFrame[number: bigint]

In [18]:

 查看 RDD 之间的继承关系可以用 explain()函数。

myDF = mySpark.range(100, 1).toDF("number").where("number % 2 = 0").filter("number % 5 = 0").sort("number").explain()

 number 为列名；where 与 filter 的本质一样；在 RDD 中，sort 有时候不显示，因为不需要排序。

== Physical Plan ==
*Sort [number#27L ASC NULLS FIRST], true, 0
+- Exchange rangepartitioning(number#27L ASC NULLS FIRST, 200)
   +- *Project [id#24L AS number#27L]
      +- *Filter (((id#24L % 2) = 0) && ((id#24L % 5) = 0))
         +- *Range (100, 1, step = 1, splits = 1)

46.4 Spark DataFrame 操作

In [19]:

 （1）读入数据，创建 Spark 数据框（DataFrame）对象。

df = mySpark.read.csv('flights.csv', header = True)

 读者可以从本书配套资源中下载文件 flights.csv。

In [20]:

 （2）显示 DataFrame 的模式信息。

df.printSchema()

Out [20]: root
 |-- year: string (nullable = true)
 |-- month: string (nullable = true)

```
|-- day: string (nullable = true)
|-- dep_time: string (nullable = true)
|-- dep_delay: string (nullable = true)
|-- arr_time: string (nullable = true)
|-- arr_delay: string (nullable = true)
|-- carrier: string (nullable = true)
|-- tailnum: string (nullable = true)
|-- flight: string (nullable = true)
|-- origin: string (nullable = true)
|-- dest: string (nullable = true)
|-- air_time: string (nullable = true)
|-- distance: string (nullable = true)
|-- hour: string (nullable = true)
|-- minute: string (nullable = true)
```

In [21]:

（3）DataFrame 对象的缓存方法：调用 cache()方法，对应的存储级别的默认取值为 MEMORY_AND_DISK。目前，Spark DataFrame 的缓存仅支持默认级别。

```
df.cache()
```

Out [21]: DataFrame[year: string, month: string, day: string, dep_time: string, dep_delay: string, arr_time: string, arr_d elay: string, carrier: string, tailnum: string, flight: string, origin: string, dest: string, air_time: string, distance: string, hour: string, minute: string]

In [22]:

（4）显示 DataFrame 的内容（取值）。DataFrame 对象的方法 show()的功能是查看数据框的内容（取值）。

```
#查看前 5 条记录
df.show(5)
```

Out [22]:
```
+----+----+----+----+----+----+----+----+----+----+----+----+----+----+----+----+
|year|month|day|dep_time|dep_delay|arr_time|arr_delay|carrier|tailnum|flight|origin|dest|air_time|distance|hou r|minute|
+----+----+----+----+----+----+----+----+----+----+----+----+----+----+----+----+
```

```
|2014| 1| 1| 1|  96| 23|  70| AS| N508AS| 145| PDX| ANC| 194| 1542| 0| 1|
|2014| 1| 1| 4|  -6|738| -23| US| N195UW| 1830| SEA| CLT| 252| 2279| 0| 4|
|2014| 1| 1| 8|  13|548|  -4| UA| N37422| 1609| PDX| IAH| 201| 1825| 0| 8|
|2014| 1| 1|28|  -2|800| -23| US| N547UW|  466| PDX| CLT| 251| 2282| 0|28|
|2014| 1| 1|34|  44|325|  43| AS| N762AS|  121| SEA| ANC| 201| 1448| 0|34|
+----+----+----+----+----+----+----+----+----+----+----+----+----+----+----+-
only showing top 5 rows
```

In [23]:

提示　（5）显示 DataFrame 的列名。通过 DataFrame 对象的 columns 属性，可以查看数据框的列名。

df.columns

Out [23]:　['year',
'month',
'day',
'dep_time',
'dep_delay',
'arr_time',
'arr_delay',
'carrier',
'tailnum',
'flight',
'origin',
'dest',
'air_time',
'distance',
'hour',
'minute']

In [24]:

提示　（6）统计 DataFrame 的行数。DataFrame 对象的方法 count() 的功能是统计数据框的行数。

df.count()

Out [24]:　52535

In [25]:

（7）选择数据框的特定列。DataFrame 对象的方法 select()用于选择数据框中的指定列。

```
spark_df_flights_selected = df.select(df['tailnum'], df['flight'],\
df['dest'], df['arr_delay'], df['dep_delay'])
```

In [26]:

查看选择数据的前三条记录。

```
spark_df_flights_selected.show(3)
```

Out [26]:
```
+-----+------+----+---------+---------+
|tailnum|flight|dest|arr_delay|dep_delay|
+-----+------+----+---------+---------+
|N508AS|   145| ANC|        7|        96|
|N195UW|  1830| CLT|      -23|       -6|
|N37422|  1609| IAH|       -4|       13|
+-----+------+----+---------+---------+
only showing top 3 rows
```

In [27]:

（8）将 DataFrame 转换为临时视图。调用 createGlobalTempView()方法

```
#该方法的参数（如'flights_view'）为临时视图的名称
df.createTempView('flights_view')
```

In [28]:

（9）重命名数据框，参见本书【46.6 DataFrame 的可视化（P440/In[37]）】。

（10）将数据转换为本地数据框对象，参见本书【46.6 DataFrame 的可视化（P440/In[38]）】。

（11）数据框的可视化，参见本书【46.6 DataFrame 的可视化（P441/In[39]）】。

46.5 SQL 编程

In [29]:
  （1）可使用 SparkSession 对象（如 Spark）的 sql() 方法执行 SQL 语句。

```
#构造一个SQL语句
sql_str = 'select dest, arr_delay from flights_view'
```

In [30]:
```
#执行SQL语句
spark_destDF = mySpark.sql(sql_str)
```

In [31]:
```
#查看查询结果的内容
spark_destDF.show(3)
```

Out [31]:
```
+----+---------+
|dest|arr_delay|
+----+---------+
| ANC|       70|
| CLT|      -23|
| IAH|       -4|
+----+---------+
only showing top 3 rows
```

In [32]:
  （2）将 Spark SQL 语句的查询结果写入磁盘。

```
# 导入模块tempfile，用于建立临时文件
# DataFrame对象的方法write.csv()将数据框保存为CSV文件
# 此处会新建一个Output_spark_destDF目录
# 并在其下存储CSV文件，类似HDFS的存储
import tempfile
tempfile.mkdtemp()
spark_destDF.write.csv("spark", mode = 'overwrite')
```

46 Spark 编程

In [33]:

 提示 （3）读取已保存的 Spark SQL 语句结果。SparkSession 对象的方法 read.csv() 将 CSV 文件读取为弹性式分布式 DataFrame。

```
dfnew = mySpark.read.csv('spark.csv')

# 查看DataFrame对象的内容（取值）
dfnew.show(3)
```

Out [33]:
```
+---+---+
|_c0|_c1|
+---+---+
|ANC| 70|
|CLT|-23|
|IAH| -4|
+---+---+
only showing top 3 rows
```

In [34]:

 提示 （4）过滤 DataFrame 的行。方法：filter()

```
jfkDF = df.filter(df['dest'] == 'JFK')
jfkDF.show(3)
```

Out [34]:
```
+----+-----+---+--------+---------+--------+---------+-------+-------+------+------+----+--------+--------+----+------+
|year|month|day|dep_time|dep_delay|arr_time|arr_delay|carrier|tailnum|flight|origin|dest|air_time|distance|hou r|minute|
+----+-----+---+--------+---------+--------+---------+-------+-------+------+------+----+--------+--------+----+------+
|2014|    1|  1|     654|       -6|    1455|      -10|     DL| N686DA|   418|   SEA| JFK|     273|    2422|   6|    54|
|2014|    1|  1|     708|       -7|    1510|      -19|     AA| N3DNAA|   236|   SEA| JFK|     281|    2422|   7|     8|
|2014|    1|  1|     708|       -2|    1453|      -20|     DL| N3772H|  2258|   PDX| JFK|     267|    2454|   7|     8|
+----+-----+---+--------+---------+--------+---------+-------+-------+------+------+----+--------+--------+----+------+
only showing top 3 rows
```

In [35]:

 提示 （5）分组统计 Spark 数据框。
方法：groupBy()，另用 agg() 实现聚合。

```python
# groupBy()接收一个列为参数，作为分组依据
# agg()接收一个字典作为参数，一个键值对对应一个列的操作
    # 键（key）表示待聚合的列的类名
    # 值（value）表示聚合使用的方法
dailyDelayDF = df.groupBy(df.day)\
.agg({'dep_delay': 'mean', 'arr_delay':'mean'})

# 使用 DataFrame 对象的方法 show()显示数据框的内容
  # 从显示结果可以看出，计算结果为"所有航班的每日平均延误起飞时间
    和每日平均延误降落
dailyDelayDF.show()
```

Out [35]:
```
+----+----+----+----+----+----+--
|day| avg(arr_delay)| avg(dep_delay)|
+----+----+----+----+----+----+--
| 7|0.0252152521525215241 5.243243243243243|
| 15| 1.0819155639571518| 4.818353236957888|
| 11| 5.749170537491706| 7.250661375661376|
| 29| 6.407451923076923| 11.32174955062912|
| 3| 5.629350893697084|11.526241799437676|
| 30| 9.433526011560694| 12.31663788140472|
| 8| 0.52455919395466| 4.5559045226130661|
| 22| -1.0817571690054912| 6.102314250913521|
| 28| -3.4050632911392404| 4.110270951480781|
| 16| 0.31582125603864736|4.29174201326100051|
| 5| 4.42015503875969| 8.199896960329731|
| 31| 5.796638655462185| 6.382229673093042|
| 18| -0.235370611183355|3.01949317738791431|
| 27| -4.354777070063694| 4.864126984126984|
| 17| 1.8664688427299703| 5.8738151658767781|
| 26| -1.524868344060854| 4.833430742255991|
| 6| 3.1785932721712538| 7.0750457596095181|
| 19| 2.8462462462462463| 7.2083832335329341|
| 23| 2.352836879432624| 6.307105108631826|
| 25| -2.385800401875418|3.4145527369826434|
+----+----+----+----+----+----+--
only showing top 20 rows
```

46.6 DataFrame 的可视化

In [36]: （6）查看数据框的模式信息。

```
dailyDelayDF.printSchema()
```

Out [36]:
```
root
 |-- day: string (nullable = true)
 |-- avg(arr_delay): double (nullable = true)
 |-- avg(dep_delay): double (nullable = true)
```

In [37]: （1）重命名 DataFrame 数据框。

```
# DataFrame对象的方法withColumnRenamed()的功能：重命名列名
# 接收两个实际参数，分别为原列名和新列名
# 需要注意的是，该方法并不会直接在原数据框上进行"就地修改"
# 而是返回另一个更新列名后的新数据框
dailyDelayDF = dailyDelayDF.withColumnRenamed('avg(arr_delay)', 'avg_arr_delay')
dailyDelayDF = dailyDelayDF.withColumnRenamed('avg(dep_delay)', 'avg_dep_delay')
dailyDelayDF.printSchema()
```

Out [37]:
```
root
 |-- day: string (nullable = true)
 |-- avg_arr_delay: double (nullable = true)
 |-- avg_dep_delay: double (nullable = true)
```

In [38]: （2）将数据转换为本地数据框。DataFrame 对象的 toPandas() 方法可将弹性分布式数据框转换为本地的 Pandas 数据框。

```
local_dailyDelay = dailyDelayDF.toPandas()
```

```
# 查看 Pandas 数据框前 10 行内容
local_dailyDelay.head(10)
```

Out [38]:

	day	avg_arr_delay	avg_dep_delay
0	7	0.025215	5.243243
1	15	1.081916	4.818353
2	11	5.749171	7.250661
3	29	6.407452	11.321750
4	3	5.629351	11.526242
5	30	9.433526	12.316638
6	8	0.524559	4.555905
7	22	−1.081757	6.102314
8	28	−3.405063	4.110271
9	16	0.315821	4.291742

In [39]:
```
%matplotlib inline
```

 提示　（3）上一行代码的含义是，设置 Matplotlib 绘图为"行内显示"，可在 Jupyter Notebook 中直接绘制图像，否则需调用 show()方法显示绘制的图像。

```
import matplotlib.pyplot as plt
# 绘制"日期-起飞"散点图
# 方法astype()的功能是强制类型转换
plt.scatter(local_dailyDelay.day.values.astype('i8'),\
            local_dailyDelay.avg_dep_delay.astype('f8'))
# 设置轴名
plt.xlabel('日期')
```

Out [39]: Text(0,0.5,'起飞延误时间')

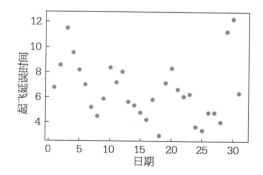

46 Spark 编程

In [40]:
```
# 绘制"日期-到达"散点图
plt.scatter(local_dailyDelay.day.values.astype('i8'),\
            local_dailyDelay.avg_arr_delay.values.astype('f8'))
# 设置X和Y轴的名称
plt.xlabel('日期')
plt.ylabel('到达延误时间')
# 绘制水平线x=0
plt.axhline(0, color = 'black', linestyle = '--', alpha = 0.5)
```

 提示 | 此处，i8 和 f8 分别代表 64 位的（8 字节）int 和 float。

Out[40]: <matplotlib.lines.Line2D at 0x251966c82b0>

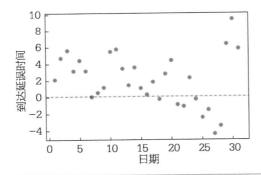

In [41]:

 提示 | 关闭 SparkSession 会话。

```
mySpark.stop()
```

46.7 Spark 机器学习

In [42]:

 提示 | 本节用 Spark 提供的机器学习库（Spark MLib）实现本书【42 统计分析】中的线性回归。

46.7.1 创建 Spark Session

In [43]:

 创建 Spark Session 并连接到 Spark 集群，参见本书【45.2 Spark Session 及其创建】。

```
from pyspark.sql import SparkSession
mySpark = SparkSession.builder\
.appName('My_LR')\
.master('local')\
.getOrCreate()
```

46.7.2 读入数据

In [44]:

 将 women.csv 文件读入 Spark 数据框对象 myDF。

```
myDF=mySpark.read.format("csv")\
.option("inferSchema", "true")\
.option("header", "true")\
.load("women.csv")
myDF
```

Out [44]:　DataFrame[_c0: int, height: int, weight: int]

In [45]:

 显示数据框 myDF 的前 5 行。

myDF.head(3)

 此处可以用方法 take() 来替代方法 head()。

Out [45]:　[Row(_c0 = 1, height = 58, weight = 115),
　　　　　　Row(_c0 = 2, height = 59, weight = 117),
　　　　　　Row(_c0 = 3, height = 60, weight = 120)]

46.7.3 数据理解

In [46]: 查看 Spark 数据框对象 myDF 的模式信息。

```
myDF.printSchema()
```

Out [46]: root
   |-- _c0: integer (nullable = true)
   |-- height: integer (nullable = true)
   |-- weight: integer (nullable = true)

In [47]: 查看 Spark 数据框 myDF 的描述性统计信息。方法 toPandas() 和 ranspose() 的功能分别为将 Spark 数据框转换为 Pandas 数据框和对 Pandas 数据框进行转置操作。

```
myDF.describe().toPandas().transpose()
```

Out [47]:

	0	1	2	3	4
summary	count	mean	stddev	min	max
_c0	15	8.0	4.47213595499958	1	15
height	15	65.0	4.47213595499958	58	72
weight	15	136.73333333333332	15.498694261437752	115	164

In [48]: 显示 Spark 数据框对象 myDF 的前 3 行。

```
myDF.take(3)
```

Out [48]: [Row(_c0 = 1, height = 58, weight = 115),
   Row(_c0 = 2, height = 59, weight = 117),
   Row(_c0 = 3, height = 60, weight = 120)]

46.7.4 数据准备

In [49]: 关于机器学习和统计分析中的数据准备方法，参见本书【42 统计分析（P370/In[1]）】和【43 机器学习（P391/In[9]）】。

注意 Spark MLib 中数据准备的基本思路与本书【42 统计分析(P370)】和【43 机器学习(P385)】中的类似,但在细节上有所区别。

In [50]:

 提示 定义特征矩阵。

```
from pyspark.ml.feature import VectorAssembler
vectorAssembler = VectorAssembler(inputCols = ['height'], outputCol = 'features')
v_myDF = vectorAssembler.transform(myDF)
v_myDF.take(3)
```

 提示 输出结果中的 DenseVector 的含义为"密集向量"。在 Spark 中,向量分为密集(Desnse)向量和稀疏(Sparse)向量。

Out [50]: [Row(_c0 = 1, height = 58, weight = 115, features = DenseVector([58.0])),
Row(_c0 = 2, height = 59, weight = 117, features = DenseVector([59.0])),
Row(_c0 = 3, height = 60, weight = 120, features = DenseVector([60.0]))]

In [51]:

 提示 提取自变量 features 和因变量 weight。

```
v_myDF = v_myDF.select(['features', 'weight'])
v_myDF.take(3)
```

Out [51]: [Row(features = DenseVector([58.0]), weight = 115),
Row(features = DenseVector([59.0]), weight = 117),
Row(features = DenseVector([60.0]), weight = 120)]

In [52]:

 提示 测试集与训练集的切分。

```
train_df = v_myDF
test_df = v_myDF
```

 思路 考虑到本例数据量过少,所以将全部数据作为训练集和测试集。如果数据量大,那么可以用方法 randomSplit() 进行训练集和测试集的切分,代码如下。

46 Spark 编程

```
# splits = v_myDF.randomSplit([0.7, 0.3])
# train_df = splits[0]
# test_df = splits[1]
```

46.7.5 模型训练

In [53]:

提示 用 Spark MLib 的 LinearRegression()函数进行简单线性回归

```
from pyspark.ml.regression import LinearRegression
myModel = LinearRegression(featuresCol = 'features', labelCol = 'weight')
```

思路 形参 featuresCol 和 labelCol 分别代表的是"特征矩阵"和"目标向量"。

```
myResults = myModel.fit(train_df)

# myModel = LinearRegression(featuresCol = 'features', labelCol = 'weight', maxIter = 10, regParam = 0.3, elasticNetParam = 0.8)
```

In [54]:
```
print("Coefficients: " + str(myResults.coefficients))
print("Intercept: " + str(myResults.intercept))
```

Coefficients: [3.4499999999999913]
Intercept: -87.51666666666614

46.7.6 模型评价

In [55]: `summary = myResults.summary`

In [56]:

提示 查看"残差"。

`summary.residuals.take(15)`

Out[56]: [Row(residuals = 2.416666666666657),
Row(residuals = 0.9666666666666401),
Row(residuals = 0.5166666666666515),
Row(residuals = 0.06666666666666288),
Row(residuals = −0.38333333333332575),
Row(residuals = −0.8333333333333144),
Row(residuals = −1.283333333333303),
Row(residuals = −1.7333333333332916),
Row(residuals = −1.1833333333332803),
Row(residuals = −1.633333333333269),
Row(residuals = −1.0833333333332575),
Row(residuals = −0.5333333333332462),
Row(residuals = 0.016666666666736774),
Row(residuals = 1.5666666666667481),
Row(residuals = 3.1166666666667595)]

In [57]:
 查看"判定系数 R 方"。

summary.r2

Out[57]: 0.9910098326857506

In [58]:
 查看"均方残差"。

summary.rootMeanSquaredError

Out [58]: 1.419702629269787

46.7.7 预测

In [59]:
```
predictions = myResults.transform(test_df)
predictions.show()
```

Out [59]:
```
+----+----+----+----+--
|features|weight| prediction|
+----+----+----+----+--
| [58.0]| 115|112.58333333333334|
| [59.0]| 117|116.03333333333336|
```

```
|[60.0]| 120|119.48333333333335|
|[61.0]| 123|122.93333333333334|
|[62.0]| 126|126.38333333333333|
|[63.0]| 129|129.83333333333331|
|[64.0]| 132| 133.2833333333333|
|[65.0]| 135| 136.7333333333333|
|[66.0]| 139|140.18333333333328|
|[67.0]| 142|143.63333333333327|
|[68.0]| 146|147.08333333333326|
|[69.0]| 150|150.53333333333325|
|[70.0]| 154|153.98333333333326|
|[71.0]| 159|157.43333333333325|
|[72.0]| 164|160.88333333333324|
+----+----+----+----+--
```

In [60]: predictions.select("prediction").show()

Out [60]:
```
+--------------+
| prediction|
+--------------+
|112.58333333333334|
|116.03333333333336|
|119.48333333333335|
|122.93333333333334|
|126.38333333333333|
|129.83333333333331|
| 133.2833333333333|
| 136.7333333333333|
|140.18333333333328|
|143.63333333333327|
|147.08333333333326|
|150.53333333333325|
|153.98333333333326|
|157.43333333333325|
|160.88333333333324|
+--------------+
```

In [61]:

 关闭 Spark Session。

mySpark.stop()

47 基于 Spark 和 MongoDB 的大数据分析

 常见疑问及解答。

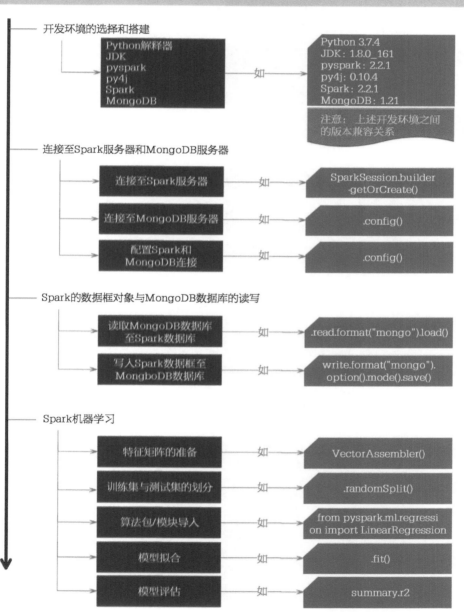

47 基于 Spark 和 MongoDB 的大数据分析

In [1]:
> **思路** 本章综合运用 Spark、MongoDB 和机器学习方法，对其中的父亲（Father）和儿子（Son）身高进行线性回归分析。
>
> **提示** 本章数据来自 Kaggle 平台，文件名为 "Pearson.txt"，数据内容源自 Karl Pearson 在 1903 年对 1078 对父子的观察，原始数据中添加了随机噪声，使身高精确到 0.1 英寸。
>
> **提示** 读者可以在本书配套资源中找到文件 Pearson.txt。

47.1 数据准备

In [2]:
> **思路** 创建 SparkSession 对象 mySpark，并连接至 Spark 服务器和 MongoDB 服务器。

```
from pyspark.sql import SparkSession

mySpark = SparkSession \
    .builder \
    .appName("myApp") \
    .config("spark.mongodb.input.uri", "mongodb://127.0.0.1/local.FSHeight") \
    .config("spark.mongodb.output.uri", "mongodb://127.0.0.1/local.FSHeight") \
    .config('spark.jars.packages','org.mongodb.spark:mongo-spark-connector_2.11:2.4.1')\
    .getOrCreate()
```

In [3]:
> **思路** 调用 PySpark 的 read() 方法将数据文件 FSHeight.txt 中的数据读入至 Spark 数据框 sparkFSHeight。

```
sparkFSHeight=mySpark.read.format("csv")\
    .option("inferSchema", "true")\
    .option("delimiter",'\t')\
    .option("header", "true")\
    .load("FSHeight.txt")
```

In [4]:

 显示 Spark 数据框 sparkFSHeight 的前 5 行。

```
sparkFSHeight.show(5)
```

Out [4]:
```
+------+----+
|Father| Son|
+------+----+
|  65.0|59.8|
|  63.3|63.2|
|  65.0|63.3|
|  65.8|62.8|
|  61.1|64.3|
+------+----+
only showing top 5 rows
```

In [5]:

 显示 Spark 数据框 sparkFSHeight 的行数。

```
sparkFSHeight.count()
```

 原始数据的总行数为 1078。

Out [5]: 1078

In [6]:

 显示 Spark 数据框 sparkFSHeight 的模式信息。

```
sparkFSHeight.printSchema()
```

47 基于 Spark 和 MongoDB 的大数据分析

```
root
 |-- Father: double (nullable = true)
 |-- Son: double (nullable = true)
```

In [7]:

 将 Spark 数据框 sparkFSHeight，以 overwrite（重写）模式存入 MongoDB 数据库 local 中的数据集 FSHeight。

```
sparkFSHeight.write.format("mongo").option("uri","mongodb://127.0.0.1/local.FSHeight").mode("overwrite").save()
```

 若采用 append（追加方式）写入数据，多次运行此行会导致数据重复写入。

In [8]:

 关闭 SparkSession。

```
mySpark.stop()
```

47.2 数据读入

In [9]:

 创建 SparkSession 对象 mySpark，并连接至 Spark 服务器和 MongoDB 服务器。

```
from pyspark.sql import SparkSession

mySpark = SparkSession \
    .builder \
    .appName("myApp") \
    .config("spark.mongodb.input.uri",  "mongodb://127.0.0.1/local.FSHeight") \
    .config("spark.mongodb.output.uri", "mongodb://127.0.0.1/local.FSHeight") \
    .config('spark.jars.packages','org.mongodb.spark:mongo-spark-connector_2.11:2.4.1')\
    .getOrCreate()
```

In [10]:
> **思路** 用 PySpark 的 read() 方法读取 MongoDB 中的数据集 FSHeight。
>
> sparkDF_FSHeight=mySpark.read.format("mongo").load()

In [11]:
> **思路** 显示数据框 sparkDF_FSHeight 的前 5 行。
>
> sparkDF_FSHeight.head(5)
>
> **提示** 可以用 take() 方法替代 head() 方法。

Out [11]: [Row(Father=65.0, Son=59.8, _id=Row(oid='5f2fad9782e983388a215c75')),
Row(Father=63.3, Son=63.2, _id=Row(oid='5f2fad9782e983388a215c76')),
Row(Father=65.0, Son=63.3, _id=Row(oid='5f2fad9782e983388a215c77')),
Row(Father=65.8, Son=62.8, _id=Row(oid='5f2fad9782e983388a215c78')),
Row(Father=61.1, Son=64.3, _id=Row(oid='5f2fad9782e983388a215c79'))]

In [12]:
> **思路** 显示数据框 sparkDF_FSHeight 的行数。
>
> sparkDF_FSHeight.count()
>
> **提示** 如果显示行数并非为 1078，需要清空 MongoDB 中的 FSHeight 数据集，删除命令为：db.FSHeight.remove({})。

Out [12]: 1078

47.3 数据理解

In [13]:
> **思路** 查看 Spark 数据框——sparkDF_FSHeight 的模式信息。
>
> sparkDF_FSHeight.printSchema()

47 基于 Spark 和 MongoDB 的大数据分析

```
root
 |-- Father: double (nullable = true)
 |-- Son: double (nullable = true)
 |-- _id: struct (nullable = true)
 |    |-- oid: string (nullable = true)
```

In [14]:

 查看 Spark 数据框——sparkDF_FSHeight 的描述性统计信息。

```
sparkDF_FSHeight.describe().toPandas().transpose()
```

 方法 toPandas() 和 ranspose() 的功能分别为将 Spark 数据框转换为 Pandas 数据框和对 Pandas 数据框进行转置操作。

Out [14]:

	0	1	2	3	4
summary	count	mean	stddev	min	max
Father	1078	67.68682745825602	2.745827077877217	59.0	75.4
Son	1078	68.68423005565862	2.8161940362006628	58.5	78.4

In [15]:

 显示 Spark 数据框——sparkDF_FSHeight 的前 10 行。

```
sparkDF_FSHeight.take(10)
```

Out [15]:
```
[Row(Father=65.0, Son=59.8, _id=Row(oid='5f2fad9782e983388a215c75')),
 Row(Father=63.3, Son=63.2, _id=Row(oid='5f2fad9782e983388a215c76')),
 Row(Father=65.0, Son=63.3, _id=Row(oid='5f2fad9782e983388a215c77')),
 Row(Father=65.8, Son=62.8, _id=Row(oid='5f2fad9782e983388a215c78')),
 Row(Father=61.1, Son=64.3, _id=Row(oid='5f2fad9782e983388a215c79')),
 Row(Father=63.0, Son=64.2, _id=Row(oid='5f2fad9782e983388a215c7a')),
 Row(Father=65.4, Son=64.1, _id=Row(oid='5f2fad9782e983388a215c7b')),
 Row(Father=64.7, Son=64.0, _id=Row(oid='5f2fad9782e983388a215c7c')),
 Row(Father=66.1, Son=64.6, _id=Row(oid='5f2fad9782e983388a215c7d')),
 Row(Father=67.0, Son=64.0, _id=Row(oid='5f2fad9782e983388a215c7e'))]
```

In [16]:

 数据可视化。

```
pandasDF_FSHeight=sparkDF_FSHeight.toPandas()
```

 调用 toPandas()方法将 Spark 数据框（sparkDF_FSHeight）转换为 Pandas 数据框（pandasDF_FSHeight）。

import matplotlib.pyplot as plt
%matplotlib inline

 绘制"日期–起飞"散点图。

plt.scatter(pandasDF_FSHeight["Son"],pandasDF_FSHeight["Father"])
plt.xlabel('Height of Sons')
plt.ylabel('Height of Fathers')

Out [16]:   Text(0, 0.5, 'Height of Fathers')

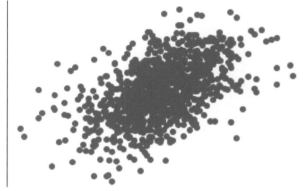

47.4 数据准备

In [17]:

 显示数据框 pandasDF_FSHeight 的前 5 行。

pandasDF_FSHeight.head()

Out [17]:

	Father	Son	_id
0	65.0	59.8	(5f2fad9782e983388a215c75,)
1	63.3	63.2	(5f2fad9782e983388a215c76,)
2	65.0	63.3	(5f2fad9782e983388a215c77,)

47 基于 Spark 和 MongoDB 的大数据分析

| 3 | 65.8 | 62.8 | (5f2fad9782e983388a215c78,) |
| 4 | 61.1 | 64.3 | (5f2fad9782e983388a215c79,) |

In [18]:

 显示数据框 pandasDF_FSHeight 的形状。

pandasDF_FSHeight.shape

Out [18]: (1078, 3)

In [19]:

 对数据内容进行 Z-score 标准化处理。

from scipy import stats
import numpy as np
z = np.abs(stats.zscore(pandasDF_FSHeight[["Son","Father"]]))
print(z)

Out [19]: [[3.1561581 0.97896716]
 [1.94829456 1.59837581]
 [1.91276917 0.97896716]
 ...
 [0.21875472 1.49866744]
 [0.21875472 1.09787361]
 [0.59832943 0.9521304]]

In [20]:

 异常值处理。

zscored_pandasDF_FSHeight = pandasDF_FSHeight[(z < 3).all(axis=1)]
zscored_pandasDF_FSHeight.shape

Out [20]: (1067, 3)

In [21]:

 将 Spark 数据框 sparkFSHeight，以 overwrite（重写）模式存入 MongoDB 数据库 local 中的数据集 zscored_sparkDF_FSHeight。

```
zscored_sparkDF_FSHeight=mySpark.createDataFrame(zscored_pandasDF_FSHeight)

zscored_sparkDF_FSHeight.write.format("mongo").option("uri","mongodb://127.0.0.1/local.ZScoredFSHeight").mode("overwrite").save()

sparkDF_FSHeight=mySpark.read.format("mongo").option("uri","mongodb://127.0.0.1/local.ZScoredFSHeight").load()

sparkDF_FSHeight.count(),len(sparkDF_FSHeight.columns)
```

Out [21]: (1067, 3)

In [22]:

 思路 定义特征矩阵。

```
from pyspark.ml.feature import VectorAssembler
```

 提示 用包 pyspark.ml.feature 的方法 VectorAssembler()，将多个列（用 inputCols 表示）合并成一个"向量列（vector column，用 outputCol 表示）。

```
vectorAssembler = VectorAssembler(inputCols = ['Father'], outputCol = 'features')

v_sparkDF_FSHeight = vectorAssembler.transform(sparkDF_FSHeight)

v_sparkDF_FSHeight.take(3)
```

Out [22]: [Row(Father=63.3, Son=63.2, _id=Row(oid='5f2fad9782e983388a215c76'), features=DenseVector([63.3])),
 Row(Father=64.6, Son=69.5, _id=Row(oid='5f2fad9782e983388a215e8e'), features=DenseVector([64.6])),
 Row(Father=65.6, Son=70.1, _id=Row(oid='5f2fad9782e983388a215cfc'), features=DenseVector([65.6]))]

47 基于 Spark 和 MongoDB 的大数据分析

In [23]:

提取自变量 features 和因变量 Son。

v_sparkDF_FSHeight = v_sparkDF_FSHeight.select(['features', 'Son'])

v_sparkDF_FSHeight.take(3)

Out [23]: [Row(features=DenseVector([63.3]), Son=63.2),
 Row(features=DenseVector([64.6]), Son=69.5),
 Row(features=DenseVector([65.6]), Son=70.1)]

In [24]:

训练集和测试集的划分。

train_DF = v_sparkDF_FSHeight

test_DF = v_sparkDF_FSHeight

提示 样本量较少，所以将训练集和测试集均设为整个数据集。

技巧
用 .randomSplit() 函数进行训练集和测试集切分，代码如下：
splits = v_sparkDF_FSHeight.randomSplit([0.7, 0.3])
train_DF = splits[0]
test_DF = splits[1]

显示测试集的行数和列数。

test_DF.count(), len(test_DF.columns)

Out [24]: (1067, 2)

47.5 模型训练

In [25]:

用 Spark MLib 的 LinearRegression() 函数进行简单线性回归。

```python
from pyspark.ml.regression import LinearRegression

myModel = LinearRegression(featuresCol = 'features', labelCol= 'Son')
```

 提示 | featuresCol 和 labelCol 分别代表的是特征矩阵和目标向量。

 思路 | 进行拟合: fit()

```python
myResults = myModel.fit(train_DF)
```

In [26]:
```python
print("Coefficients: " + str(myResults.coefficients))
print("Intercept: " + str(myResults.intercept))
```

Out [26]: Coefficients: [0.4918272114633771]
Intercept: 35.39422233085258

47.6 模型评价

In [27]:

 思路 | 生成预测模型的统计信息对象。

```python
summary = myResults.summary
```

In [28]:

 思路 | 查看残差。

```python
summary.residuals.show()
```

 注意 | 用 show() 方法才能显示残差信息。

47 基于 Spark 和 MongoDB 的大数据分析

Out [28]:
```
+------------------+
|         residuals|
+------------------+
|-3.3268848164843376|
| 2.3337398086132595|
|    2.44191259714988|
|-4.062991075972093|
|  2.19272987600354|
|-0.6203465856550423|
|-4.956452845142792|
| 0.9450357226379396|
| 2.4943644337108566|
| 0.6845570874669278|
|-1.1448649512649212|
|-0.964625633679418|
| 0.8466702803452648|
| 2.6943644337108594|
|-2.1793366530453255|
|-0.9662601913867377|
| 4.704171779954805|
| 1.9566236165157989|
|-3.4597219605574594|
| 1.38455708746691641|
+------------------+
only showing top 20 rows
```

In [29]: 查看判断系数 R 方。

summary.r2

Out [29]: 0.2502440487185249

In [30]: 查看均方残差。

summary.rootMeanSquaredError

Out [30]: 2.325165606265826

47.7 模型应用

In [31]:

 进行预测。

predictions = myResults.transform(test_DF)

 显示预测结果。

predictions.show()

Out [31]:
```
+------+----+------------------+
|features|Son|        prediction|
+------+----+------------------+
|  [63.3]|63.2|66.52688481648434|
|  [64.6]|69.5|67.16626019138674|
|  [65.6]|70.1|67.65808740285011|
|  [65.0]|63.3|67.36299107597209|
|  [65.7]|69.9|67.70727012399647|
|  [64.1]|66.3|66.92034658565504|
|  [65.8]|62.8|67.75645284514279|
|  [72.1]|71.8|70.85496427736206|
|  [65.9]|70.3|67.80563556628914|
|  [64.7]|67.9|67.21544291253308|
|  [61.1]|64.3|65.44486495126492|
|  [64.8]|66.3|67.26462563367942|
|  [72.3]|71.8|70.95332971965473|
|  [65.9]|70.5|67.80563556628914|
|  [63.0]|64.2|66.37933665304533|
|  [64.6]|66.2|67.16626019138674|
|  [67.1]|73.1|68.39582822004519|
|  [67.4]|70.5| 68.54337638348421|
|  [65.4]|64.1|67.55972196055745|
```

47 基于 Spark 和 MongoDB 的大数据分析

```
|    [64.7]|68.6|67.21544291253308|
+-----+--+-----------------+
only showing top 20 rows
```

In [32]:

 显示预测结果中的 prediction 列的值。

```
predictions.select("prediction").show()
```

Out [32]:
```
+-----------------+
|       prediction|
+-----------------+
|66.52688481648434|
|67.16626019138674|
|67.65808740285011|
|67.36299107597209|
|67.70727012399647|
|66.92034658565504|
|67.75645284514279|
|70.85496427736206|
|67.80563556628914|
|67.21544291253308|
|65.44486495126492|
|67.26462563367942|
|70.95332971965473|
|67.80563556628914|
|66.37933665304533|
|67.16626019138674|
|68.39582822004519|
| 68.5433763834842|
|67.55972196055745|
|67.21544291253308|
+-----------------+
only showing top 20 rows
```

In [33]:

 关闭 Spark Session 对象。

```
mySpark.stop()
```

第七篇

继续学习

Python 初学者常见错误及纠正方法
Python 数据分析和数据科学面试题
继续学习本书内容的推荐资源

48 Python 初学者常见错误及纠正方法

48.1 NameError: name 'xxxx' is not defined

In [1]:
```
#原因分析
    #Python中用户的变量、常量和函数都需要"先定义",否则报此类错误
#纠正方法
    #(1)如果是自定义变量、常量、函数、类等,需要补充其定义代码,参
见本书【5 变量(P28)】
    #(2)如果是模块函数,需要导入(import)所属模块的命名空间,参
见本书【22 模块函数(P141)】
```

In [2]:
```
#例如
x1
    #报错信息:NameError: name 'x1' is not define
    #原因分析:x1 未定义
```

Out [2]:
```
NameError Traceback (most recent call last)
<ipython-input-2-e72d8db43c78> in <module>()
      1 #例如
----> 2 x1
      3     #报错:NameError: name 'x1' is not define
      4     #原因分析:x1 未定义

NameError: name 'x1' is not defined
```

In [3]:
```
#纠正建议
x1 = 1
x1
```

Out [3]: 1

In [4]:
```
#对于初学者而言,此类错误的出现往往是由变量名的拼写错误或大小写不一
致导致的,如:
x2 = 2
X2
```

48 Python 初学者常见错误及纠正方法

In [4]:
```
#报错信息：NameError: name 'X2' is not defined
#原因分析：Python 区分大小写，x2 已定义，但未定义 X2
```

Out [4]:
```
---------------------------------------------------------
NameError Traceback (most recent call last)
<ipython-input-4-44905fb30189> in <module>()
      1 #对于初学者而言，此类错误的出现往往是由变量名的拼写错误或大小写混用导致的，如：
      2 x2 = 2
----> 3 X2
      4 #报错：NameError: name 'X2' is not defined
      5 #原因分析：Python 区分大小写，x2 已定义，但未定义 X2

NameError: name 'X2' is not defined
```

In [5]:
```
#纠正建议
x2 = 2
x2
```

Out [5]: 2

In [6]:
```
#再如
np.arange(1,10)    #报错 NameError: name 'np' is not defined
```

Out [6]:
```
---------------------------------------------------------
NameError Traceback (most recent call last)
<ipython-input-6-82f0ae180462> in <module>()
      1 #再如
----> 2 np.arange(1,10)    #报错 NameError: name 'np' is not defined

NameError: name 'np' is not defined
```

In [7]:
```
#纠正方法
    #导入模块numpy
import numpy as np
np.arange(1,10)
```

Out [7]: array([1, 2, 3, 4, 5, 6, 7, 8, 9])

48.2 IndentationError: unexpected indent

In [8]:
```
#原因分析
    #缩进错误
```

```
#纠正方法
    #纠正缩进方式，参见本书【6.4 复合语句】
```

In [9]:
```
#例如：
x = 1
    x
```

Out [9]:
```
  File "<ipython-input-9-c15b4f81a9cb>", line 3
    x
    ^
IndentationError: unexpected indent
```

In [10]:
```
#纠正建议：纠正缩进方式，如：
x = 1
x
```

Out [10]: 1

48.3 SyntaxError: invalid character in identifier

In [11]:
```
#原因分析
    #输入了中文标点符号。注意：Python 语法中，所有标点符号均为英文
状态/格式
#纠正方法
    #改为英文标点符号
```

In [12]:
```
#例如
x＝（1，2，3）    #此处圆括号和逗号为中文格式
```

Out [12]:
```
  File "<ipython-input-12-a482482883f7>", line 2
    x＝（1，2，3）    #此处圆括号和逗号为中文格式
      ^
SyntaxError: invalid character in identifier
```

In [13]:
```
#纠正建议
    #将逗号和圆括号分别改为英文格式
x=(1,2,3)
```

48.4 TypeError: 'XXXX' object does not support item assignment

In [14]:
```
#原因分析
    #Python 的对象分为"可变对象"和"不可变对象"
    #不允许对"不可变对象"进行修改/赋值操作
    #参见本书【15 元组】
```

In [15]:
```
#例如
x = (1,2,3)      #x为元组
x[0] = 100
x
```

Out [15]:
```
TypeError Traceback (most recent call last)
<ipython-input-15-f4670a5be85d> in <module>()
      1 #例如
      2 x = (1,2,3)      #x 为元组
----> 3 x[0] = 100
      4 x

TypeError: 'tuple' object does not support item assignment
```

In [16]:
```
#纠正建议：将不可变对象（此处为元组）改为可变数据类型（如列表），如
x1 = [1,2,3]      #将元组改为列表
x1[0] = 100
x1
```

Out [16]: [100, 2, 3]

48.5 TypeError: unsupported operand type(s) for XXXX

In [17]:
```
#原因分析
    #Python属于强类型语言，不支持自动类型转换
    #参见本书【5.3 Python 是强类型语言（P30/In[3]）】
#纠正方法
    #进行强制类型转换
```

In [18]:
```
#例如
1 + "20"
```

Out [18]:
```
TypeError Traceback (most recent call last)
<ipython-input-18-516797a0d119> in <module>()
      1 #例如
----> 2 1+"20"

TypeError: unsupported operand type(s) for +: 'int' and 'str'
```

In [19]:
```
#纠正建议：进行"强制类型转换"，如
1 + int("20")
```

Out[19]:   21

48.6 IndexError: list index out of range

In [20]:
```
#原因分析
    #序列的下标超出了边界
    #参见本书【17 序列（P112/In[2]）】
#纠正方法
    #改为有效下标
```

In [21]:
```
#例如
a=[1,2,3]
a[3]
```

Out [21]:
```
IndexError Traceback (most recent call last)
<ipython-input-21-41ab3f41928d> in <module>()
      1 #例如
      2 a = [1,2,3]
----> 3 a[3]

IndexError: list index out of range
```

In [22]:
```
#纠正建议
a = [1,2,3]
a[2]          #注意Python下标从0开始
```

Out[22]:   3

48.7 TypeError: type() takes XXXX arguments

In [23]:
```
#原因分析
    #函数调用时给出的实参与形参不一致
#纠正方法
    #查看对应函数的帮助信息
```

In [24]:
```
#例如
type("test1","test2")
```

Out [24]:
```
---------------------------------------------------
TypeError Traceback (most recent call last)
<ipython-input-24-ff8bf67ff8d7> in <module>()
      1 #例如
----> 2 type("test1","test2")

TypeError: type() takes 1 or 3 arguments
```

In [25]:
```
#系统显示的帮助信息如下：
# Init signature: type(self, /, *args, **kwargs)
# Docstring:
# type(object_or_name, bases, dict)
# type(object) -> the object's type
# type(name, bases, dict) -> a new type
# Type: type
```

In [26]:
```
#纠正方法
type("test1"),type("test2")
```

Out[26]: (str, str)

48.8 SyntaxError: unexpected EOF while parsing

In [27]:
```
#原因分析
    #语句未结束或表达式不完整
```

In [28]:
```
#纠正方法
    #修改为完整语句或表达式
```

```
#例如
myList = [11,12,13,14,15,16,17,18,19]
myList[1
```

Out[28]:
```
  File "<ipython-input-28-231b8b6f3305>", line 3
    myList[1
            ^
SyntaxError: unexpected EOF while parsing
```

In [29]:
```
#纠正方法：补充缺失部分 "]"
myList=[11,12,13,14,15,16,17,18,19]
myList[1]
```

Out[29]: 12

48.9 ModuleNotFoundError: No module named XXXX

In [30]:
```
#原因分析
    #尚未下载和安装该包
#纠正方法
    #在 CMD 窗口中输入 pip install 或 conda install 来下载和安装该包
```

In [31]:
```
#例如
import tensorflow
```

Out[31]:
```
---------------------------------------------------------------
ModuleNotFoundError  Traceback (most recent call last)
<ipython-input-31-031745f846ac> in <module>()
      1 #例如
----> 2 import tensorflow

ModuleNotFoundError: No module named 'tensorflow'
```

In [32]:
```
#纠正建议：先用 PIP 或 Conda 下载或安装该包
    #在 CMD 窗口中输入 pip install tensorflow
```

48.10 TypeError: 'list' object is not callable

In [33]:
```
#原因分析
    #对应方法无法调用
    #通常,将"属性"误用为"方法"时,报此类错误
#纠正方法
    #将方法改为属性
```

In [34]:
```
#例如
import keyword
keyword.kwlist()
```

Out [34]:
```
TypeError        Traceback (most recent call last)
<ipython-input-34-b6f5e3e93bbe> in <module>()
      1 #例如
      2 import keyword
----> 3 keyword.kwlist()

TypeError: 'list' object is not callable
```

In [35]:
```
#纠正方法:将 kwlist()方法改为属性 kwlist
import keyword
keyword.kwlist
```

Out[35]:
```
['False',
 'None',
 'True',
 'and',
 'as',
 'assert',
 'break',
 'class',
 'continue',
 'def','del',
 'elif','else',
 'except',
 'finally',
```

```
'for',
'from',
'global',
'if',
'import',
'in','is',
'lambda',
'nonlocal',
'not','or',
'pass',
'raise',
'return',
'try',
'while',
'with',
'yield']
```

48.11 SyntaxError: invalid syntax

In [36]:
```
#原因分析
    #语法错误
#纠正方法
    #检查语法是否有问题
```

In [37]:
```
#例如
i = 1
j = 2
k = 3
print(i;j;k)
```

 提示：print()函数中把应写","之处错误地写成了";"。

Out [37]:
```
  File "<ipython-input-37-c7428686d33c>", line 5
    print(i;j;k)
            ^
SyntaxError: invalid syntax
```

48 Python初学者常见错误及纠正方法

In [38]:
```
#纠正建议：将分号改为逗号
i = 1
j = 2
k = 3
print(i,j,k)
```

Out [38]: 1 2 3

48.12 AttributeError:XXXX object has no attribute XXXX

In [39]:
```
#原因分析
    #该对象中无此属性
#纠正方法
    #查看对象的帮助信息或定义信息，给出正确的属性名
```

In [40]:
```
#例如
myTuple = 1,3,5,7,2
myTuple.sort()
```

 提示　报错原因：对象 myTuple 对应的类中没有成员方法 sort()。

Out [40]:
```
---------------------------------------------------------------
AttributeError            Traceback (most recent call last)
<ipython-input-40-8e13b5e95490> in <module>()
      1 #例如
      2 myTuple=1,3,5,7,2
----> 3 myTuple.sort()
      4 #提示：报错原因：对象 myTuple 对应的类中没有成员方法 sort()

AttributeError: 'tuple' object has no attribute 'sort'
```

In [41]:
```
#纠正方法
myTuple = 1,3,5,7,2
print(sorted(myTuple))
```

> 提示 | 改为 Python 内置函数 sorted()。

Out [41]: [1, 2, 3, 5, 7]

48.13 TypeError: XXXX object is not an iterator

In [42]:
```
#原因分析
    #所访问的对象并非为"迭代器"
```
> 提示 | 参见本书【25 迭代器与生成器（P165）】。
```
#纠正方法
    #改变访问方法或转换为迭代器后再访问
```

In [43]:
```
#例如
myList = [1,2,3,4,5]
next(myList)
```
> 提示 | Python 中的列表（List）是可迭代的对象，而不是迭代器。

Out [43]:
```
---------------------------------------------------------------
TypeError Traceback (most recent call last)
<ipython-input-43-bfe8be7bca79> in <module>()
      1 #例如
      2 myList = [1,2,3,4,5]
----> 3 next(myList)
      4 #提示：Python 中的列表（List）是可迭代的对象，而不是迭代器

TypeError: 'list' object is not an iterator
```

In [44]:
```
#纠正方法：先将可迭代对象转换为迭代器
myIterator = iter(myList)
print(next(myIterator))
```

48 Python 初学者常见错误及纠正方法

```
print(next(myIterator))
print(next(myIterator))
```

 提示 | 参见本书【25 迭代器与生成器（P165）】。

Out [44]: 1
2
3

48.14 FileNotFoundError: File XXXX does not exist

In [45]:
```
#原因分析
    #文件XXXX未找到，原因是：可能写错了文件名，或者目标文件不在
"当前工作目录"中
#纠正方法
    #将文件事先放在"当前工作目录"中，并检查文件名和文件路径
```

In [46]:
```
#例如
from pandas import read_csv
data = read_csv('bc_data1.csv')
data.head(2)
```

 提示 | 'bc_data.csv'并未事先放在"当前工作目录"中，即文件路径不正确。

Out [46]:
```
---------------------------------------------------------
FileNotFoundError Traceback (most recent call last)
<ipython-input-46-4558920878b3> in <module>()
      1 #例如
      2 from pandas import read_csv
----> 3 data = read_csv('bc_data1.csv')
      4 data.head(2)
      5 #提示：'bc_data.csv'并未事先放在"当前工作目录"中，即文件路径不
```

正确

C:\Anaconda\lib\site-packages\pandas\io\parsers.py in parser_f(filepath_or_buffer, sep, delimiter, header, names, index_col, usecols, squeeze, prefix, mangle_dupe_cols, dtype, engine, converters, true_values, false_values, skipinitialspace, skiprows, nrows, na_values, keep_default_na, na_filter, verbose, skip_blank_lines, parse_dates, infer_datetime_format, keep_date_col, date_parser, dayfirst, iterator, chunksize, compression, thousands, decimal, lineterminator, quotechar, quoting, escapechar, comment, encoding, dialect, tupleize_cols, error_bad_lines, warn_bad_lines, skipfooter, skip_footer, doublequote, delim_whitespace, as_recarray, compact_ints, use_unsigned, low_memory, buffer_lines, memory_map, float_precision)
    707   skip_blank_lines=skip_blank_lines)
    708
--> 709   return _read(filepath_or_buffer, kwds)
    710
    711 parser_f.__name__ = name

C:\Anaconda\lib\site-packages\pandas\io\parsers.py in _read(filepath_or_buffer, kwds)
    447
    448   # Create the parser.
--> 449   parser = TextFileReader(filepath_or_buffer, **kwds)
    450
    451   if chunksize or iterator:

C:\Anaconda\lib\site-packages\pandas\io\parsers.py in __init__(self, f, engine, **kwds)
    816   self.options['has_index_names'] = kwds['has_index_names']
    817
--> 818   self._make_engine(self.engine)
    819
    820   def close(self):

C:\Anaconda\lib\site-packages\pandas\io\parsers.py in _make_engine(self, engine)
   1047   def _make_engine(self, engine='c'):
   1048   if engine == 'c':
-> 1049   self._engine = CParserWrapper(self.f, **self.options)

```
1050            else:
1051                if engine == 'python':

C:\Anaconda\lib\site-packages\pandas\io\parsers.py in __init__(self, src, **kwds)
    1693            kwds['allow_leading_cols'] = self.index_col is not False
    1694
-> 1695    self._reader = parsers.TextReader(src, **kwds)
    1696
    1697            # XXX
```

pandas/_libs/parsers.pyx in pandas._libs.parsers.TextReader.__cinit__()

pandas/_libs/parsers.pyx in pandas._libs.parsers.TextReader._setup_parser_source()

FileNotFoundError: File b'bc_data1.csv' does not exist

In [47]:
```python
#纠正方法：事先将文件放在"当前工作目录中"
from pandas import read_csv
data = read_csv('bc_data.csv')
data.head(2)
```

Out [47]:

	id	diag-nosis	radius_mean	texture_mean	perimeter_mean	area_mean	smoothne
0	842302	M	17.99	10.38	122.8	1001.0	0.11840
1	842517	M	20.57	17.77	132.9	1326.0	0.08474

2 rows × 32 columns

48.15 IndexError: too many indices for array

In [48]:
```python
#原因分析
    #索引（下标）的维度超出了其该对象的定义中给出的维数
#纠正方法
    #改用Fancy Indexing的方法
```

In [49]:
```python
#例如
import numpy as np
```

```
myArray = np.array(range(1,11))
myArray[1,3,6]
```

> **提示** 从 myArray 的定义看，myArray 只有 1 维，但是在其调用 myArray[1,3,6]中给出的维度数量为 3。

Out [49]:
```
----------------------------------------------------------------
IndexError Traceback (most recent call last)
<ipython-input-50-0172781e9f33> in <module>()
      2 import numpy as np
      3 myArray = np.array(range(1,11))
----> 4 myArray[1,3,6]
      5 #提示：从 myArray 的定义看，myArray 只有 1 维，但是在其调用
myArray[1,3,6]中给出的维度数量为 3
IndexError: too many indices for array
```

In [50]:
```
#纠正方法：改用切片的方法
    #关于切片，参见本书【36 数组（P245）】中对 Fancy Indexing 的解释。
myArray[[1,3,6]]
```

Out [50]: `array([2, 4, 7])`

48.16 TypeError: Required argument XXXX not found

In [51]:
```
#原因分析
    #在方法/函数的调用中没有给出必选形式参数对应的实际参数的值
#纠正方法
    #查看该方法/函数的说明文档，给出必选形式参数的值
```

In [52]:
```
#例如
import datetime as dt
dt.datetime(month = 3, day = 3, second = 59)
```

> **提示** 从 dt.datetime 的说明文档看，year、month、day 为必选，其他为可选。

48　Python 初学者常见错误及纠正方法

Out[52]:
```
TypeError  Traceback (most recent call last)
<ipython-input-53-22bfc45c99c6> in <module>()
      1 #例如
      2 import datetime as dt
----> 3 dt.datetime(month = 3, day = 3, second = 59)
      4 #提示：从 dt.datetime 的说明文档看，year、month、day 为必选，其他为可选

TypeError: Required argument 'year' (pos 1) not found
```

In[53]:
```
#纠正方法：给出必选形式参数的值，参见本书【23　自定义函数（P150/In[12]）】
dt.datetime(year = 2018, month = 3, day = 3, second = 59)
```

Out[53]:　datetime.datetime(2018, 3, 3, 0, 0, 59)

48.17　TypeError: an XXXX is required (got type YYYY)

In[54]:
```
#原因分析
    #对象的类型有误
    #错误提示的含义为：此处需要的是 XXXX 类型的对象，而给出的是 YYYY 类型
#纠正方法
    #检查拼写是否有误，如果不是拼写错误，那么建议进行强制类型转换或换成另一个函数
```

In[55]:
```
#例如
dt.datetime("3th of July, 2018")
```

提示　报错信息：此处需要的是一个整型（int），但读入的是字符串（str）。

Out[55]:
```
TypeError  Traceback (most recent call last)
<ipython-input-56-9889e1d0c642> in <module>()
      1 #例如
----> 2 dt.datetime("3th of July, 2018")
      3 #【提示】报错误信息：此处需要的是一个整数（int），但读入的是字符
```

串（str）

TypeError: an integer is required (got type str)

In [56]:
```
#纠正方法：进行强制类型转换或用正确的函数，参见本书【4.3 转换数据类型的方法（P22/In[12]）】
from dateutil import parser
MyDate = parser.parse("3th of July,2018")
```

Out [56]: datetime.datetime(2018, 7, 3, 0, 0)

48.18 ValueError: Wrong number of items passed XXXX,placement implies YYYY

In [57]:
```
#原因分析
    #代码中给出的items个数是XXX，但解释器需要的个数为YYYY
#纠正方法
    #按解释器要求调整 items 的个数
```

In [58]:
```
#例如
import pandas as pd
s3 = pd.Series([1,2,3,4,5], index = ["a","b","c"])    #注意：出错
s3
```

Out [58]:
```
---------------------------------------------------------
ValueError Traceback (most recent call last)
<ipython-input-59-034044b18868> in <module>()
      1 #例如
      2 import pandas as pd
----> 3 s3 = pd.Series([1,2,3,4,5], index = ["a","b","c"])    #注意：出错
      4
      5 s3

C:\Anaconda\lib\site-packages\pandas\core\series.py in __init__(self, data, index, dtype, name, copy, fastpath)
    264                                        raise_cast_failure = True)
--> 266             data = SingleBlockManager(data, index, fastpath = True)
    267
```

48 Python初学者常见错误及纠正方法

```
    268                 generic.NDFrame__init__(self, data, fastpath = True)

C:\Anaconda\lib\site-packages\pandas\core\internals.py in __init__(self, block, axis, do_integrity_check, fastpath)
   4400         if not isinstance(block, Block):
   4401             block = make_block(block, placement=slice(0, len(axis)), ndim=1,
-> 4402                                fastpath=True)
   4403
   4404         self.blocks =[block]

C:\Anaconda\lib\site-packages\pandas\core\internals.py in make_block(values, placement, klass, ndim, dtype, fastpath)
   2955                      placement = placement, dtype = dtype)
   2956
-> 2957     return klass(values, ndim = ndim, fastpath = fastpath, placement = placement)
   2958
   2959 # TODO: flexible with index = None and/or items = None

C:\Anaconda\lib\site-packages\pandas\core\internals.py in __init__(self, values, placement, ndim, fastpath)
    118             raise ValueError('Wrong number of items passed %d, placement '
    119                              'implies %d' % (len(self.values),
--> 120                              len(self.mgr_locs)))
    121
    122     @property

ValueError: Wrong number of items passed 5, placement implies 3
```

In [59]:
```
#纠正方法：按错误提示进行修改items个数，参见本书【37 Series（P265）】
s3 = pd.Series([1,2,3], index = ["a","b","c"])
s3
```

Out [59]: a   1
           b   2
           c   3
           dtype: int64

49 Python 数据分析和数据科学面试题

 提示 | 本章为互联网收集到的岗位面试题，用于检查和提升读者的学习效果、解决实际问题和就业能力。

（1）本章仅针对无法在本书中直接找到答案的或容易答错的问题给出了提示。
（2）考虑到个别考题的原题为使用 Python 2 编写和国外语言文化的特殊性，在确保原题性质不变的情况下，进行了一定的优化。
（3）本题库并非全部为作者原创内容，主要来自于网络招聘信息，如：

[1] Data Science in Python Interview Questions and Answers[OL]. https://www.mytectra.com/interview-question/data-science-inpython-interview-questions-and-answers/

[2] 25 Python Questions & Answers for Data Science Interviews[OL]. https://www.springpeople.com/blog/25-python-questionsanswers-for-data-science-interviews/

[3] 100 Data Science in Python Interview Questions and Answers for 2018[OL]. https://www.dezyre.com/article/100-data-science-in-pythoninterview-questions-and-answers-for-2018/188

[4] Top 50 Python Interview Questions You Must Prepare In 2018[OL]. https://www.edureka.co/blog/interview-questions/pythoninterview-questions/

[5] 8 Important Python Interview Questions and Answers[OL]. https://www.datablogger.com/2017/11/17/important-python-interview-questions-and-answers/

[6] Top Python Interview Questions And Answers[OL]. https://intellipaat.com/interview-question/python-interviewquestions/

[7] 15 Essential Python Interview Questions[OL]. https://www.codementor.io/sheena/essential-pythoninterview-questions-du107ozr6

[8] 11 Essential Python Interview Questions[OL]. https://www.toptal.com/python/interview-questions

[9] Top 40 Python Interview Questions & Answers[OL]. https://career.guru99.com/top-25-python-interview-questions/

[10] Python Interview Questions[OL]. https://mindmajix.com/python-interviewquestions

[11] What are commonly asked Python interview questions[OL]. https://www.quora.com/What-are-commonly-asked-Pythoninterview-questions

[12] 30 Essential Python Interview Questions You Should Know.[OL]. http://www.techbeamers.com/python-interview-questionsprogrammers/

49 Python 数据分析和数据科学面试题

1. 你曾参与的数据分析/数据科学相关的 Python 项目有哪些？你负责的是哪些工作？遇到了哪些挑战？如何解决的？有哪些收获与体会？
2. 你用过（或正在用）哪些 Python 包或模块？如何用的？有什么体会？
3. 在 Python 编程中遇到问题时，你通常如何解决的？
4. 你用（或学）Python 多久了？做过哪些项目？你自己最引以为豪的项目是哪一个？
5. 你是否有 Github 等开源社区的账户？有没有自己发起的项目？
6. 目前有哪些著名的 Python 开源项目？你参与或关注过哪一个？
7. 为什么在数据分析或数据科学中不用 C、Java、C#，而用 Python 或 R 语言？
8. 你认为，在数据分析或数据科学实践中，R 语言和 Python 哪一个更好？为什么？
9. 到底什么是 Python？请与其他语言进行对比分析，给出 Python 的主要特征。
10. 数据科学中常用的 Python 包或模块有哪些？
11. 用 Python 做数据分析与用 Python（或 C、Java、C#等）做软件开发之间的区别是什么？
12. 与 C、Java、C#比较，Python 有哪些特征？你对 Python 的哪些特征满意？对哪些特征不满意？
13. 你最喜欢的机器学习算法是什么？如何用 Python 实现它？
14. 你处理过的最大数据集是什么？你是如何处理的？最后结果是什么？是否用了 Python？如果没有，能否改为 Python？
15. 在数据科学或数据分析相关的项目中，你是否用过 A/B 测试？涉及 Python 吗？
16. Python 在数据分析或数据科学中常用算法和模型有哪些？
17. 为什么说"Python 是强类型语言"？
18. 为什么说"Python 是可执行的伪代码"？
19. 为什么说"Python 是解释型语言"？
20. 为什么说"Python 是动态类型语言"？
21. 为什么说"Python 是自带电池（Batteries Included）的"？
22. 如何理解"Python 中的鸭子类型编程"？
23. 如何理解"Python 中一切皆为对象"？
24. 什么是 Python 中的模块？如何调用？请举例说明。
25. Pandas 是什么？主要应用场景是什么？
26. Python 内置数据类型有哪些？
27. set 与 frozenset 的区别是什么？
28. 基于 Python 如何进行探索性数据分析？
29. 简要说明 Pandas 中的 DataFrame。
30. Python 可变对象和不可变对象是指什么？请分别列举说明。
31. Python 基础语法中是否有数组的概念？Python 中数组的主要功能是如何实现的？
32. 模块函数的导入方法有几种？不同导入方法之间有什么区别？
33. 在 Python 中用 lambda 定义一个函数和用 def 定义一个函数有什么区别？
34. Python 中的 yield 语句是什么意思？
35. Python 函数定义中，带有一个星号（*）和带有两个星号（**）的形式参数之间的区别是什么？

36. 在数据科学实践中，Python 字典类型的主要应用场景是什么？
37. 在 Python 中，模块和包的区别与联系是什么？请举例说明。
38. 什么是 NumPy？在数据科学中的主要功能是什么？
39. 什么是 Matplotlib？在数据科学中的主要功能是什么？
40. 什么是可迭代对象？什么是迭代器？
41. 迭代器与生成器的区别和联系是什么？
42. 什么是闭包？如何定义？如何调用？
43. 什么是变量的搜索路径？如何查看变量的搜索路径？
44. 如何查看和修改模块的搜索路径？
45. 什么是当前工作目录？如何查看它？如何修改它？
46. 如何列出一个 Python 模块中的函数名清单？
47. 在 Python 中，序列是指什么？常见序列有哪些？序列有共性特征？
48. 什么是 PIP 工具？如何用？
49. PIP 和 Conda 的区别是什么？
50. Python 编程中常用的 IDE 有哪些？你在数据分析/数据科学中用的是哪一个？
51. 在 Python 中，何种情况下用字典，而不用列表？
52. Python 中的代码块（或复合语句）是如何定义的？
53. Python 的函数如何返回一个值？
54. Python 中是否有 do-while 语句？
55. break 与 continue 语句之间的区别是什么？请举例说明。
56. 什么是 ufunc？请结合 Python 编程进行说明。
57. 列表的 extend 和 append 有什么区别？
58. 列表的 resize 和 reshape 有什么区别？
59. 在 Python 中，help()和 dir()的区别是什么？
60. 在 Python 中，如何实现三元组运算符的功能？
61. 在 Python 中，元组和列表的区别是什么？
62. 在数据分析或数据科学中，通常用来实现机器学习的包有哪些？
63. 如何用 Python 实现一个简单的逻辑回归模型？
64. 在用 Python 进行数据可视化时，你喜欢用 Seaborn 还是 MatplotLib？
65. Pandas 的 Series 与包含一列的数据框（DataFrame）有什么区别？
66. 如何对数据框（DataFrame）进行排序操作？请用代码说明。
67. 在 Python 中，如何将某个对象中的重复值进行重复过滤？
68. Python 中的负索引是什么意思？如何用？举例说明。
69. 如何在 NumPy 数组中获得 N 个最大值的索引？
70. 通过调整什么参数就可以提升随机森林的预测效果？
71. 举例说明 Python 中的内置函数 zip()。
72. 举例说明 Python 中的内置函数 filter()。
73. 举例说明 Python 中的内置函数 enumerate()。

49 Python 数据分析和数据科学面试题

74. 举例说明 np.split() 的用法。
75. 在 Python 中，如何进行切片操作？请举例说明。
76. 在 Python 中，如何将整数转换为字符串？
77. Local 变量与 Global 变量之间的区别是什么？请举例说明。
78. 在 Python 中，"//" 与 "\\" 的区别是什么？
79. 能否定义一个包含多个数据类型的 DataFrame？如果可以，如何定义？如果不能，请说明原因。
80. 能否直接用 Pandas 包绘制直方图？如果能，对应代码是什么？如果不能，原因是什么？
81. 在 Python 中，为什么内置函数的实现不用 Python，而是用 C 语言等其他语言？
82. lambda 函数是什么？请举例说明。
83. 在 Python 中，docString 指的是什么？
84. 在 Python 中，列表、集合和元组的区别是什么？
85. 在 Python 中，全局变量、局部变量、非局部变量分别用什么关键字来标识？
86. 在 Python 中，如何区分一个函数的必选参数和可选参数？
87. 在 Python 中，如何区分一个函数的位置参数和关键词参数？
88. 在 Python 中，什么是强制命名参数？请举例说明。
89. 在 Python 中，装饰器是什么？请解释其用法。
90. 在 Python 面向对象编程中，self 和 cls 的区别是什么？
91. 在 Python 面向对象编程中，new 和 init 的区别是什么？
92. 什么是 iPython 的魔术命令？请举例说明在数据科学中的应用。
93. Pandas 如何表示缺失值？
94. 在 Python 机器学习中，如何进行训练集和测试集的划分？
95. 在数据科学中，为什么用 NumPy 数组，而不用 Python 基础语法中的 list？
96. 在 Python 中如何进行程序调试？（提示：除了本书给出的方法，通常还可以用 Pylint 和 Pychecker。）
97. 如何判断一个 NumPy 数组是否为空？
98. 在数据规整化处理中，通常用什么 Python 包？
99. 在 NumPy 中，sort() 和 argsort() 的区别是什么？
100. 哪一个（些）Python 包建立在 Matplotlib 和 Pandas 上，而且实现了更好的可视化效果？
101. 在 Python 中，如何生成一个随机数？如何生成一个随机数组？
102. 如何复制一个 Python 对象？深拷贝与浅拷贝之间的区别是什么？
103. 什么是 PEP8？（提示：PEP8 即 Style Guide for Python Code。）
104. Python 的 Monkey Patch（猴子补丁）有哪些特征？（提示：支持在运行时对已有的代码进行修改。）
105. __init__.py 是用来做什么的？
106. TensorFlow 是什么？你在 Python 编程中用过它吗？如何用的？
107. Python 中的变量命名规范是什么？
108. 解释 Python 中的 os 模块。

109. 什么是列表推导式？
110. range()与np.arange()的区别是什么？
111. 在Python中，如何混洗一个列表？（提示：random模块中的shuffle函数。）
112. 在Python中，pass语句是什么意思？
113. 在Python中，函数参数传递采用值传递还是地址传递？
114. 在Python中，常见的*args与**kwargs之间的区别是什么？
115. 在Python中，装饰器是什么？请举例说明其使用方法。
116. 如何用Python进行自然语言处理？有哪些难点？
117. 如何用MatplotLib进行数据可视化？
118. np.sort()与np.argsort()有什么区别？
119. 如何判断一个DataFrame（数据框）是否为空？
120. lambda函数中可否定义内部变量？
121. 在Python中如何进行模式匹配？（提示：用re模块。）
122. 在Scikit-learn中如何导入决策树分类器？（提示：from sklearn.tree import DecisionTreeClassifier。）
123. 在Python数据分析或数据科学实践中，很多函数中经常包括参数axis = 0或axis = 1，这是什么意思？
124. 在Python中，如何反向遍历（从最后一个元素到第一个元素的方向读出）一个列表的内容？
125. 在Python数据分析或数据科学实践中，很多函数中经常包括参数inplace = True或inplace = False，它们是什么意思？
126. 在Python中如何进行内存管理？
127. 如何对DataFrame创建多级索引？
128. 什么是假正率？通常如何计算它的？
129. 在Python数据分析或数据科学实践中，如何查看数据的形状？如何更改数据对象的形状？
130. 在Python中如何获得URL？（提示：如www.abcd.org的获取方式为urllib2.urlopen（www.abcd.org）或requests.get（www.abcd.org）。）
131. 基于统计学的数据科学实践的基本步骤是什么？
132. 基于机器学习的数据科学实践的基本步骤是什么？
133. 你在Python编程中常用的包有哪些？你负责（或参与）开发过什么包？开发Python包的基本步骤是什么？
134. 什么是JSON？在Python中，如何将JSON转换为Pandas的DataFrame？
135. Spark Session是什么？如何创建？
136. Spark DataFrame与Pandas DataFrame的区别是什么？如何在二者之间进行转换？
137. 在Spark的Python编程中如何调用SQL语句？
138. 用Python编写网络爬虫的基本思路是什么？
139. 什么是面向对象编程？Python中如何实现面向对象编程？
140. 什么是异常处理？Python中如何进行异常处理？

49 Python 数据分析和数据科学面试题

141. 举例说明 Python 中的继承机制。
142. @classmethod、@staticmethod 与@property 的区别是什么？
143. 在 Python 中，如何实现继承与重写父类中的方法？
144. 在 Python 中，".py"文件与".pyc"文件的区别是什么？
145. 如何用 Python 调用 MySQL 数据库？（提示：MySQLdb 模块。）
146. 请解释 Python 的垃圾回收机制。（提示：请读者自行了解。）
147. 编写一个函数，用来判断单词是否为回文。
148. 什么是 Pylab？
149. 编写一个函数，其输入为两个已排序的向量，返回值为一个合并成重新排序的向量。
150. 以下代码的输出结果为_____。
    ```
    def multipliers ():
        return [lambda x : i * x for i in range (4)]
    print [m (2) for m in multipliers ()]
    ```
    （提示：[6, 6, 6, 6]）
151. 以下代码的输出结果为_____。
    ```
    word = 'aeioubcdfg'
    print(word [:3] + word [3:])
    ```
152. 以下代码的输出结果为_____。
    ```
    myString = "I Love Python"
    for myChar in myString:
        if myChar == " ":
            continue
    print(myChar)
    ```
    （提示：n ）
153. 以下代码的输出结果为_____。
    ```
    def multipliers():
        return [lambda x : i * x for i in range(4)]
    print([m(2) for m in multipliers()])
    ```
154. 以下代码的输出结果为_____。
    ```
    list = ['a', 'b', 'c', 'd', 'e']
    print(list[10:])
    ```
    （提示：不会报错，输出结果为：[]。）
155. 以下代码的输出结果为_____。
    ```
    list = [ [ ] ] * 5
    print(list)
    list[0].append(10)
    print(list)
    list[1].append(20)
    ```

```
       print(list)
       list.append(30)
       print(list)
       (提示：
         [[], [], [], [], []]
         [[10], [10], [10], [10], [10]]
         [[10, 20], [10, 20], [10, 20], [10, 20], [10, 20]]
         [[10, 20], [10, 20], [10, 20], [10, 20], [10, 20], 30] )
156. 以下代码的输出结果为_____。
       def foo (I = []):
           i.append (1)
           return i
       foo ()
       foo ()
157. 以下代码的输出结果为_____。
       def f(x, l = []):
           for i in range(x):
               l.append(i*i)
           print(l)
       f(2)
       f(3, [3, 2, 1])
       f(3)
       (提示：
         [0, 1]
         [3, 2, 1, 0, 1, 4]
         [0, 1, 0, 1, 4] )
158. 以下代码的输出结果为_____。
       class Parent(object):
           x = 1
       class Child1(Parent):
           pass
       class Child2(Parent):
           pass
       print(Parent.x, Child1.x, Child2.x)
       Child1.x = 2
       print (Parent.x, Child1.x, Child2.x)
       Parent.x = 3
       print (Parent.x, Child1.x, Child2.x)
```

（提示：
　1 1 1
　1 2 1
　3 2 3　）

159. 以下代码的输出结果为____。
list= ['a', 'e', 'i', 'o', 'u']
print(list [8:])
（提示：空 list[]。）

160. DataFrame（数据框）可以接收哪些类型的数据？
　　A. ndarray　　　　　　　　　B. Series
　　C. DataFrame　　　　　　　　D. 以上都可以

161. Series 是否被视为"一种带有标签的一维数组，而且可以包含任何类型的数据的结构"？
　　A. 可以
　　B. 不可以

162. 如果数据为 ndarray 型，那么其 index 是否必须与数据等长？
　　A. 是
　　B. 否

163. 在 Python 中，函数返回值是否有 void 类型？
　　A. 是
　　B. 否

164. 以下几个选项中，可以生成 Python 字典类型对象的是____。
　　A. d = {}　　　　　　　　　　B. d = {"john":40," peter":45}
　　C. d = {40:"john",45:"peter"}　D. d = {40:"john",45:"50"}
　　（提示：A、B、C、D。）

165. 以下选项中，哪一个（些）选项为 Python 无效（错误）代码？
　　A. abc = 1,000,000　　　　　　B. abc=100020003000
　　C. a,b,c = 1000, 2000, 3000　　D. a_b_c = 1,000,000
　　（提示：B。）

166. 已知 list1 = [2, 33, 222, 14, 25]，那么 list1[-1]为____。
　　A. 错误　　　　　　　　　　　B. None
　　C. 25　　　　　　　　　　　　D. 2

167. 以下代码的输出结果为____。
f = None
for i in range (5):
　　with open("data.txt", "w") as f:
　　　　if i>2:
　　　　　　break
print(f.closed)

（提示：True。）

168. 如果"以写数据为目的"打开文件"C:\scores.txt"，以下选项中哪一个为最优方案？
    A. outfile = open("C:\scores.txt", "r")
    B. outfile = open("C:\\scores.txt", "w")
    C. outfile = open(file = "C:\scores.txt", "r")
    D. outfile = open(file = "C:\\scores.txt", "o")

169. Python 中的 try-except-else 的 else 部分什么时候被执行？
    A. 任何时候   B. 当一个异常出现时
    C. 当任何异常都不发生时   D. 在 except 部分中指定的异常发生时

170. 以下代码的输出结果为____。

    class A(object):
        def go(self):
            print("go A go!")
        def stop(self):
            print("stop A stop!")
        def pause(self):
            raise Exception("Not Implemented")
    class B(A):
        def go(self):
            super(B, self).go()
            print("go B go!")
    class C(A):
        def go(self):
            super(C, self).go()
            print("go C go!")
        def stop(self):
            super(C, self).stop()
            print("stop C stop!")
    class D(B,C):
        def go(self):
            super(D, self).go()
            print("go D go!")
        def stop(self):
            super(D, self).stop()
            print("stop D stop!")
        def pause(self):
            print("wait D wait!")
    class E(B,C): pass

```
a = A()
b = B()
c = C()
d = D()
e = E()
# specify output from here onwards
a.go()
b.go()
c.go()
d.go()
e.go()
a.stop()
b.stop()
c.stop()
d.stop()
e.stop()

a.pause()
b.pause()
c.pause()
d.pause()
e.pause()
```

171. 如果

    ```
    import time
    str = '21/01/2017'
    datetime_value = time.strptime(str,date_format)
    ```
    那么，在 date_format 的位置上应该写哪一个选项中的代码？
    A. "%d/%m/%y"   B. "%D/%M/%Y"
    C. "%d/%M/%y"   D. "%d/%m/%Y"

172. 已知

    ```
    train_set = np.array([1, 2, 3])
    test_set = np.array([[0, 1, 2], [1, 2, 3]])
    ```
    那么，如何将 resulting_set 取值为 [[1, 2, 3], [0, 1, 2], [1, 2, 3]]？
    A. resulting_set = train_set.append(test_set)
    B. resulting_set = np.concatenate([train_set, test_set])
    C. resulting_set = np.vstack([train_set, test_set])
    D. 以上都不是。
    (提示：B。)

173. 以下代码的输出结果为_____。
    ```
    def extendList(val, list=[]):
    ```

```
            list.append(val)
            return list
    list1 = extendList(10)
    list2 = extendList(123,[])
    list3 = extendList('a')
    print("list1 = %s",list1)
    print("list2 = %s",list2)
    print("list3 = %s",list3)
```
（提示：
```
        list1 = %s [10, 'a']
        list2 = %s [123]
        list3 = %s [10, 'a'] )
```

174. 以下代码的输出结果是_____。
```
    A0 = dict(zip(('a','b','c','d','e'),(1,2,3,4,5)))
    A1 = range(10)
    A2 = sorted([i for i in A1 if i in A0])
    A3 = sorted([A0[s] for s in A0])
    A4 =[i for i in A1 if i in A3]
    A5 = {i:i*i for i in A1}
    A6 = [[i,i*i] for i in A1]
    print(A0,A1,A2,A3,A4,A5,A6)
```
（提示：
{'a': 1, 'b': 2, 'c': 3, 'd': 4, 'e': 5} range(0, 10) [] [1, 2, 3, 4, 5] [1, 2, 3, 4, 5]
{0: 0, 1: 1, 2: 4, 3: 9, 4: 16, 5: 25, 6: 36, 7: 49, 8: 64, 9: 81} [[0, 0], [1, 1],
[2, 4], [3, 9], [4, 16], [5, 25], [6, 36], [7, 49], [8, 64], [9, 81]])

50 继续学习本书内容的推荐资源

50.1 重要网站

[1] Python 官网：有很多权威资料，如 Python Tutorial 等，URL：Python.org。
[2] Pypi 官网：Python 包索引(Python packages index)，可以看到每个包的帮助文档，URL：https://pypi.org/project/pip/。
[3] LearnPython.org：学习 Python 的著名网站。
[4] DataCamp：用 Python 学习数据科学的著名网站：https://www.datacamp.com/。
[5] PyData：Python 数据分析的著名社区，URL：https://groups.google.com/forum/#!forum/pydata。
[6] pystatsmodels：Python 统计分析的著名讨论社区，URL：https://groups.google.com/forum/#!forum/pystatsmodels。
[7] Python cheat sheet：一图讲解 Python 知识，URL：https://ehmatthes.github.io/pcc/cheatsheets/README.html。
[8] Python2 和 Python3 的区别：https://wiki.python.org/moin/Python2orPython3。
[9] PEP 8,Python 写代码规范(Style Guide for Python Code),URL:https://www.python.org/dev/peps/pep-0008/。
[10] Kaggle：有很多数据科学和 Python 相关的竞赛、数据集等，URL：https://www.kaggle.com。
[11] Python Weekly:Python 每周一报，报告内容包括新闻、文章、新版本发布、工作岗位等，建议订阅。URL：https://www.pythonweekly.com/。
[12] GitHub：有很多 Python 开源项目，可以参见《Top 20 Python AI and Machine Learning projects on Github》等相关研究报告，建议读者多参与开源项目。当然，也有 Python 之父 Guido van Rossum 等大牛发起的开源项目。
[13] Stack Overflow：虽然不是只针对 Python 和数据科学的网站，却是本书作者最喜欢的网站之一，URL：https://stackoverflow.com/。

50.2 推荐图书

[1]《Python Data Science Handbook》（Jake VanderPlas）：一本 Python 和数据科学相结合的好书，推荐认真阅读。

[2] 《Python for Data Analysis: Data Wrangling with Pandas, NumPy, and IPython》：另一本 Python 和数据科学相结合的好书，推荐认真阅读。
[3] 《Hands-on Machine Learning with Scikit-learn & Tensorflow》：一本关于"如何用 Python 和机器学习做数据科学项目"的好书，推荐认真阅读。
[4] 《Thoughtful Machine Learning with Python: A Test-Driven Approach》：一本关于"如何用 Python 做机器学习任务"的好书，推荐认真阅读。
[5] 《Spark-The Definitive Guide》（Bill Chambers 和 Matei Zaharia）：含有用 Python 讲解 Spark 的内容，值得阅读。
[6] 《数据科学》（朝乐门）：国内第一本系统讲解数据科学理论的重要专著。另，《数据科学理论与实践》（朝乐门）中介绍了数据科学的核心理论、关键技术和典型实践，值得深入阅读。

除了上述图书，学习 Python 基础语法，还可以参考如下经典图书：《Python for Everyone》（Cay Horstmann Rance Necaise）、《Learning Python》（Mark Lutz）、《Python Pocket Reference》（Mark Lutz）和《Fluent Python》（Luciano Ramalho）。上述图书有很多版本，建议读者参考最新版本的图书。

50.3 常用模块与工具包

[1] 基础库：Pandas, Numpy, Scipy。
[2] 绘图及可视化：Matplotlib, Seaborn, Bokeh, Basemap, Plotly, NetworkX。
[3] 机器学习：SciKit-Learn, TensorFlow, Theano, Keras。
[4] 统计分析：Statsmodels。
[5] 自然语言处理、数据挖掘及其他：NLTK, Gensim, Scrapy, Pattern。

50.4 常用统计模型

[1] 广义线性模型（是多数监督机器学习方法的基础，如逻辑回归和 Tweedie 回归）
[2] 时间序列方法（ARIMA、SSA、基于机器学习的方法）
[3] 结构方程建模（针对潜变量之间关系进行建模）
[4] 因子分析（调查设计和验证的探索型分析）
[5] 功效分析/试验设计（特别是基于仿真的试验设计，以避免分析过度）
[6] 非参数检验（MCMC）
[7] K 均值聚类

[8] 贝叶斯方法（朴素贝叶斯、贝叶斯模型平均/Bayesian model averaging、贝叶斯适应性试验/Bayesian adaptive trials 等）

[9] 惩罚性回归模型（弹性网络/Elastic Net、LASSO、LARS 等）以及对通用模型（SVM、XGBoost 等）加罚分，这对于预测变量多于观测值的数据集很有用，在基因组学和社会科学研究中较为常用）

[10] 样条模型/Spline-based models（MARS 等）：主要用于流程建模

[11] 马尔可夫链和随机过程（时间序列建模和预测建模的替代方法）

[12] 缺失数据插补方法及其假设（missFores、MICE 等）

[13] 生存分析/Survival analysis（主要特点是考虑了每个观测出现某结局的时间长短）

[14] 混合建模/Mixture modeling

[15] 统计推断和组群测试（A/B 测试以及用于营销活动的更复杂的方法）

此外，建议读者根据自己所属领域重点学习面向特定领域的专用模型。

50.5 核心机器学习算法

[1] 回归/分类树

[2] 降维（PCA、MDS、tSNE 等）

[3] 经典的前馈神经网络

[4] Bagging ensembles 方法（随机森林、KNN 回归集成）

[5] Boostingensembles 方法（梯度提升、XGBoost 算法）

[6] 参数调整或设计方案的优化算法（遗传算法、量子启发式演化算法、模拟退火/simulated annealing、粒子群优化/particle-swarm optimization）

[7] 拓扑数据分析工具，特别适用于小样本量的无监督学习（持续同调/persistent homology、Morse-Smale 聚类、Mapper 等）

[8] 深度学习架构（通用深度学习架构）

[9] 用于局部建模的 KNN 方法（回归、分类）

[10] 基于梯度的优化方法/Gradient-based optimization methods

[11] 网络度量/Network metrics 和算法（中心度量、跳数、多样性、熵、拉普拉斯算子、疫情传播/epidemic spread、谱聚类/spectral clustering）

[12] 深层架构中的卷积和池化层/pooling layers（特别适用于计算机视觉和图像分类模型）

[13] 分层聚类（与 K 均值聚类和拓扑数据分析工具相关）

[14] 贝叶斯网络（路径挖掘/pathway mining）

[15] 复杂性和动态系统（与微分方程有关）

此外，建议读者根据自己所属领域还可能需要与自然语言处理、计算机视觉相关算法以及面向特定领域的专用算法。

50.6 继续学习数据科学的建议路线图

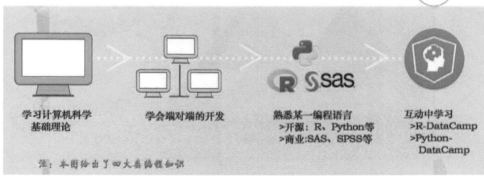

50 继续学习本书内容的推荐资源

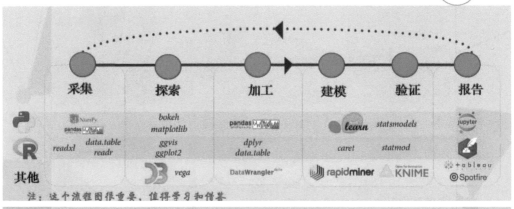

6 成长、协作与学习

积极参加相关竞赛

挑战自己并
拓宽自己的技能

与数据科学家合作

与数据科学爱好者
保持联系

要有自己的宠物项目

拥有自己的代表作，
并提升故事化描述能力

培育数据科学家精神

理论联系实际

注：这一点很多初学者都忽视了……

7 彻底浸泡自己

实习	集训	职场工作
★☆☆	★★☆	★★★
初学者	中级水平	高级水平

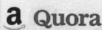

注：还记得著名的"泡菜理论"吗？

8 参与社区

跟踪最新动态

加入一个或多个社区

跟进相关信息源

争取为社区做出贡献

跟随行业中的领军人物

注：数据科学家的必经之路

50　继续学习本书内容的推荐资源

References 参考文献
http://businessoverbroadway.com/when-does-education-level-matter-in-data-science
http://drewconway.com/zia/2013/3/26/the-data-science-venn-diagram
https://www.datacamp.com
http://www.forbes.com/sites/bernardmarr/2015/06/22/spark-or-hadoop-which-is-the-best-big-data-framework/#27ed36e2532c
https://www.oreilly.com/ideas/seven-reasons-why-i-like-spark
http://www.prooffreader.com/2016/09/battle-of-data-science-venn-diagrams.html

翻译与注解人：朝乐门　微信公众号：数据科学DataScience

（注：上图源自 DataCamp，由本书作者进行翻译和注解。）

开源课程建设行动。

本书配套资源（数据文件、源代码、勘误信息等）存放的 URL 为 https://github.com/LemenChao/PythonFromDAtoDS。若无法访问 gitHub，请通过 chaolemen@ruc.edu.cn 联系本书作者。

本书根据作者朝乐门老师提出的"开源课程建设行动倡议"，为高校教师提供开源社区，提供教师上课所需的新开课申请表、课程简介、教学大纲、PPT、习题、教师参考书目等资源，请教师通过发送邮件至 chaolemen@ruc.edu.cn 索取教学资源，并欢迎参与开源课程建设行动。

参考文献

[1] 朝乐门. 数据科学. 北京：清华大学出版社, 2016.

[2] 朝乐门. 数据科学理论与实践（第二版）. 北京：清华大学出版社, 2019.

[3] VanderPlas J. Python data science handbook: essential tools for working with data. "O'Reilly Media, Inc, 2016.

[4] McKinney W. Python for data analysis: Data wrangling with Pandas, NumPy, and IPython. O'Reilly Media, Inc, 2012.

[5] Matei Zaharia, Bill Chambers. Spark: The Definitive Guide Big Data Processing Made Simple. O'Reilly Media, Inc, 2018.

[6] Géron A. Hands-on machine learning with Scikit-Learn and TensorFlow: concepts, tools, and techniques to build intelligent systems. O'Reilly Media, Inc, 2017.

[7] Lutz M. Learning Python: Powerful Object-Oriented Programming. O'Reilly Media, Inc, 2013.

[8] Garrett Grolemund, Hadley Wickham. R for Data Science. O'Reilly Media, Inc, 2017.

[9] Ramalho L. Fluent Python: clear, concise, and effective programming. O'Reilly Media, Inc, 2015.

反侵权盗版声明

电子工业出版社依法对本作品享有专有出版权。任何未经权利人书面许可，复制、销售或通过信息网络传播本作品的行为；歪曲、篡改、剽窃本作品的行为，均违反《中华人民共和国著作权法》，其行为人应承担相应的民事责任和行政责任，构成犯罪的，将被依法追究刑事责任。

为了维护市场秩序，保护权利人的合法权益，我社将依法查处和打击侵权盗版的单位和个人。欢迎社会各界人士积极举报侵权盗版行为，本社将奖励举报有功人员，并保证举报人的信息不被泄露。

举报电话：（010）88254396；（010）88258888
传　　真：（010）88254397
E-mail：　dbqq@phei.com.cn
通信地址：北京市万寿路173信箱
　　　　　电子工业出版社总编办公室
邮　　编：100036